FOUNDATIONS OF
BUSINESS TELECOMMUNICATIONS MANAGEMENT

APPROACHES TO INFORMATION TECHNOLOGY

Series Editor
Thomas F. Carbery, *Strathclyde Business School*
University of Strathclyde
Glasgow, Scotland

Foundations of Business Information Systems
Andrew Doswell

Foundations of Business Telecommunications Management
Kenneth C. Grover

Humanizing Technology: Computers in Community Use and
   Adult Education
Elisabeth Gerver

A Continuation Order Plan is available for this series. A continuation order will bring delivery of each new volume immediately upon publication. Volumes are billed only upon actual shipment. For further information please contact the publisher.

FOUNDATIONS OF
BUSINESS TELECOMMUNICATIONS MANAGEMENT

KENNETH C. GROVER
Telecommunications Consultant

PLENUM PRESS • NEW YORK AND LONDON

Library of Congress Cataloging in Publication Data

Grover, Kenneth C.
   Foundations of business telecommunications management.

   (Approaches to information technology)
   Bibliography: p.
   Includes index.
   1. Business — Communication systems. 2. Telecommunication systems. 3. Telephone in business. I. Title.
HF5541.T4G76   1986   384   86-5063
   ISBN-13: 978-0-306-42249-2   e-ISBN-13: 978-1-4613-2193-4
   DOI: 10.1007/978-1-4613-2193-4

To the memory of my son David

FOREWORD

Information technology is about more than computers.

Thus, it was a recurring—and rather infuriating—aspect of the early discussions on information technology that those who participated tended either to ignore or to severely understate the role in information technology of telecommunications. This very fine book by Ken Grover goes a long way toward correcting that misconception. However important the computer and computer-based equipment might be, the role of telecommunications equipment has also been and continues to be significant. Moreover, as the author brings out, it is going to be even more important. As this enthralling story unfolds the reader will find him or herself continually remarking that there cannot be more—but again and again, there is.

Those who are already of the world of telecommunications will, on reading this work, be proud of their colleague. Those who are already of the world of computers will learn a great deal and, it is to be hoped, will in future be fairer toward telecommunications than they have been in the past. Those who are new to the world of information technology will sally forth better balanced than most.

Thomas F. Carbery
Strathclyde Business School
University of Strathclyde
Glasgow

PREFACE

For almost a century since the invention of the telephone by Alexander Graham Bell there was no real change in telecommunications as a means for two people to converse at a distance. Moreover, telephone users had very little choice in the installations they had or the services available, having virtually always to take what the telephone companies, usually exercising a monopoly, were prepared to provide, and to accept the operating standards they decided were appropriate. Businesses generally valued the telephone as a means of communication, but, with good surface mail and the telegraph for occasional special needs, were not critically dependent on the telephone.

With the rapid and continuing advance in telecommunications technology over the last twenty years or so, there has come a dramatic growth in the dependence of industry and commerce on telecommunications. Reliance is now placed on telecommunications not only for the immediate conveyance of information by the spoken word, but also for the conveyance of data and other nonvoice intelligence. The operational and financial viability of industry and commerce now depend critically on telecommunications.

It is not surprising that the monopoly exercised by the long established telecommunications companies for so many years has been challenged. In many countries telecommunications users are being given increasing freedom to decide for themselves what installations and services best meet their needs, and have increasing influence on operating standards. In some countries companies are being allowed to develop networks and systems in competition with the existing ones. At the same time, manufacturers are entering the telecommunications and allied markets for the first time to compete in the supply of terminals and systems.

Senior managers and members of boards of major telecommunications users, of competing network and service providers, and of telecommunications manufacturers, are now facing for the first time major investment decisions on which the future commercial viability of their businesses critically depend. This book seeks to inform them, going no further into the technical aspects than is essential, and to give them an insight into the complexities of the issues with which they must cope to reach sound decisions.

AUTHOR'S NOTE

The views expressed in this book are based on the author's experience over more than forty years in the British Post Office and later, British Telecommunications. They do not necessarily represent the views of those organizations or of any other telecommunications company. The author remains very conscious, however, of the influence which the many colleagues with whom he was privileged to work have had in forming those views.

CONTENTS

PART I: THE GROWTH AND DEVELOPMENT
        OF TELECOMMUNICATIONS

Chapter 1: INTRODUCTION 3

Chapter 2: HISTORICAL PERSPECTIVE 7

Chapter 3: TECHNOLOGICAL REVOLUTION 11

Chapter 4: THE INTERFACE BETWEEN THE USER AND
        THE SYSTEM 15

Chapter 5: DATA COMMUNICATIONS 19

PART II: TELECOMMUNICATIONS SYSTEMS

Chapter 6: TELECOMMUNICATIONS TERMINALS 25
    Voice Terminals 25
    Message Terminals 29
    Data Terminals 32
    Other Terminals 34
    Terminals–Summary 37

Chapter 7: LOCAL TELECOMMUNICATION NETWORKS 39
    Internal Networks 39
    Local Networks Transmission 41
    External Local Networks 42

Chapter 8: TELECOMMUNICATIONS SWITCHING 45
    Methods of Selection 45
    Active and Passive Switching Systems 47
    Common User and Dedicated Switching Systems 49

Chapter 9: MAIN TELECOMMUNICATIONS NETWORKS 51
    Main Networks Construction 52
    Value Added Network Services and Leasing Operations 55
    Main Networks Summary 56

PART III: TELECOMMUNICATIONS STANDARDS
         OF SERVICE

Chapter 10: STANDARDS OF SERVICE 59

Chapter 11: SPEED OF INSTALLMENT 63
    Distribution of Lines 63
    Installation of Terminals 65

Chapter 12: STANDARDS OF INSTALLATION 67

Chapter 13: AVAILABILITY OF COMMUNICATIONS 71
    Switchboards 72
    Switching Equipment 73
    Main Networks 76
    Overall Standard 76

Chapter 14: SPEED OF CONNECTION 79

Chapter 15: QUALITY OF CONNECTION 81

Chapter 16: RELIABILITY 83

Chapter 17: SPEED OF REPAIR 85

Chapter 18: SECURITY OF COMMUNICATIONS 89

Chapter 19: ADEQUACY OF SUPPORT SERVICES 91
    Number Information 91
    Assistance 94
    Public Pay Phones 95

Chapter 20: USER RELATIONS 97

Chapter 21: CHARGING ACCURACY 99

Chapter 22: ENHANCEMENT AND DEVELOPMENT 101

PART IV: TELECOMMUNICATIONS PLANNING

Chapter 23: THE NATURE OF
            TELECOMMUNICATIONS PLANNING 105

Chapter 24: STRATEGIC PLANNING 107

Chapter 25: DEVELOPMENT AND
            PROCUREMENT PLANNING 109
Determination of Current Usage of Communications 109
Forecasts of Future Requirements 111
Identification of Technical Options 113
Identification of External Factors Impinging on
Telecommunciations Planning 115
Financial Appraisal of Options 116
Preparation of a Development and Procurement Plan 117
Analysis of a Development and Procurement Plan 118

Chapter 26: PREPARATION OF TELECOMMUNICATIONS
            UTILIZATION PLANS 121

PART V: TELECOMMUNICATIONS
        OPERATIONAL MANAGEMENT

Chapter 27: RESPONSIBILITY 125

Chapter 28: ACHIEVEMENT OF STANDARDS OF SERVICE 127
Operator Services 127
Dialed Services 133
Installation Repair Service 137
Provision of Service 139
Support Services 142

Chapter 29: RESOURCE MANAGEMENT 145
Equipment Utilization 145
Manpower Productivity 147
Industrial Relations 150
Financial Performance 152
Accountability 153

PART VI: FREEDOM OF CHOICE

Chapter 30: MONOPOLY OF TELECOMMUNICATIONS 159

Chapter 31: REGULATION OF TELECOMMUNICATIONS 163
    Historical Background 163
    Liberalization 169
    Competing Networks 171
    Cellular Radio Networks 172
    Value Added Network Services 173
    The Financial Implications of Liberalization 174
    Privatization 177

PART VII: CONCLUSION

Chapter 32: FUTURE TRENDS 181

Chapter 33: PERSPECTIVE 183

APPENDICES 185
    United Kingdom Growth of Telephones in Service 185
    United Kingdom Growth of Inland Telephone Calls 185
    United Kingdom Investment in Local Telephone Exchanges 186
    United Kingdom Investment in
    Telecommunications Distribution 187
    United Kingdom Telecommunications
    Annual Income and Expenditure 187
    United Kingdom Net Book Value of Fixed Assets 188
    Telecommunications Operational Expenditure 188
INDEX 189

1

THE GROWTH AND
DEVELOPMENT OF
TELECOMMUNICATIONS

1

INTRODUCTION

Throughout history there have been periods when man and the society in which he lives have entered sudden, rapid, and far-reaching periods of development and change. The Renaissance was such a period when the spread of learning and the exploration of the world brought people into closer contact and resulted in dramatic social and commercial changes. The significance of these changes was recognized by few at the time, but in retrospect is seen to have changed the course of mankind.

The development of the printing press reinforced the effects of the Renaissance. The knowledge which man was acquiring could now be more widely published and communicated to others, accelerating the spread of learning, which led in turn to the Industrial Revolution. Again over a period of less than a century the industrial, commercial, and social structures of countries were changed and the way of life altered. In retrospect the effects of the Industrial Revolution are seen as not always raising the quality of life or being desirable. The new industries depended on the employment of growing numbers of people in often dangerous, unhealthy working conditions for long hours on work which was boring and repetitious. (It is sometimes overlooked, on the other hand, that the life of the Eighteenth Century agricultural worker was hard, and rural life was not as idyllic as sometimes described by fiction writers.)

Few politicians and others exercising authority at the time recognized, understood, or were concerned that the course of the Industrial Revolution was not always improving the quality of life of mankind at large. Even less did people in general understand, recognize, or challenge the undesirable consequences of the Industrial Revolution and seek to influence its course. It was not until almost a century had passed that there grew a wider realization that not all change was to the general good and that some control should be exercised over the course of events. Sadly, in large part it was too late, and the Industrial Revolution continued to run its course. In bringing large numbers of people to work and live together, however, it fostered in turn the development of communications. The procurement of the raw materials used in the new factories and the marketing of the growing output of products had to be organized. The rising urban population had to be fed and their basic needs met. Growth in communications was inevitable, albeit at first met by postal and other message services.

The Industrial Revolution in Western society has certainly now lost its momentum. The heavy engineering industries on which it was based are in a sharp decline from which it appears doubtful that they will recover. We have now entered, however, another period of industrial, commercial, and social change as significant as the Industrial Revolution or even the Renaissance itself, and which in turn is accelerating the growth in communications. The technological revolution, based on the application of electronics, is developing faster and changing the way of life of mankind more radically than the Industrial Revolution did in its time. Largely as the result of television, radio, and other forms of mass communication there is now a wider understanding that these changes are in progress and there is a growing pressure to influence and control their course.

The application of information technology in particular and the realization of its benefits are already lagging behind the development of new and more advanced systems. New electronic based products and services could potentially raise industrial and commercial efficiency and contribute to improving the quality of life but, wastefully, are not being put into place or utilized extensively enough so that this potential can be realized; the gap between innovation and application could widen. This is attributable to a large extent to a lack of real understanding of the implications of the new technology and how to cope with the management of technological change on the part of those having to make major investment decisions.

Information technology is concerned with the capture, transmission and reception, storage, processing, and retrieval of all forms of data, voice and nonvoice. (Speech is being increasingly converted and carried electronically as a form of data, together with data from computer, telemetry, and other electronic based systems.) Communications equipment required for information technology systems can account for a major part of investment and ongoing operating costs. The purpose of this book is to develop an understanding of how those requirements can best be met and most economically utilized. It aims to provide a framework on which the critical stategic and investment decisions involved can be based. It examines the operational characteristics and potential of various ways of providing communications. It covers both main network and local distribution systems, and dedicated and common user systems.

The senior management and boards of directors of companies must now have a greater understanding of the nature of telecommunications and take a greater responsibility for the management of the systems they use. They can no longer delegate responsibility for communications to professional middle managers: The commercial future of their firms is now too heavily dependent on communications systems to allow such devolution of responsibility. They now have major investment options to exercise and decisions to make, as increasingly the commerical viability and security of their companies come to depend on telecommunications.

With the rapidly growing importance of telecommunications to political, economic, and social life, governments have ruled that the virtual monopoly of public communications exercised by state, quasi-governmental, and private companies should be countered and controlled. This has led to the statutory protection of the monopolizing companies being withdrawn and to competitors being encouraged to offer products and services. In place of the old statutory protection the operation of communications networks are being regulated in various ways in an attempt to ensure that the major companies do not, through the sheer size of their operations, continue to dominate this key service industry.

Whatever views are held about the ownership and regulation of communications, it is essential that their nature and the issues are understood. This book seeks to inform and contribute to the understanding of senior managers faced with the responsibility of investing heavily in new telecommunications, and also those who have a wider interest in the nature and complexity of the issues underlying the administration and management of telecommunications. The student of telecommunications and the professional telecommunications engineer may also find that it gives a different, but necessary, perspective to their interest.

The application of information technology raises many moral and social issues. It is not the purpose of this book to discuss or argue such issues, as those are for political consideration. The author remains convinced that given a proper understanding of the technology and a recognition of how and in what ways its application should be regulated, information technology can contribute to raising the quality of life and any unacceptable consequences can be minimized, if not avoided completely. The senior managers on whom responsibility rests for the major investment decisions involved also carry, indirectly, a responsibility for ensuring that the moral and social implications of their decisions are recognized and taken into account. Failure to do so could lead to such political, industrial, and social disruption as to jeopardize the successful application of the technology.

2

HISTORICAL PERSPECTIVE

Since his earliest days, man has sought to communicate over distances further than the human voice would carry. He has used fire, in the form of beacons, to warn of an impending enemy attack or to convey some other message of national importance. The birth of a royal baby was often made known throughout the kingdom by lighting beacons on hilltops, the sight of one being the signal to light the next in the chain, and thus to relay the news of the birth from hilltop to hilltop throughout the land. (Chains of fire beacons continue to be used today from time to time, to celebrate some special royal occasion.) Smoke has been used to convey messages and by interrupting its flow (an early form of coded signal) the message could be varied according to the information to be conveyed. Other forms of visual communication used at various times include flags; signalling gantries (a form of semaphore); and the heliograph (reflecting the light of the sun by an arrangement of mirrors). (Naval ships still use signalling lamps, and it is interesting to reflect that the latest communications technology has returned to light as a communications medium, using optical fibers to carry the light over distances beyond visual range.)

Man has used drums, another early form of coded signal, and by varying the drum beat he has varied the information conveyed. He has used bugles, again varying the sound to convey different messages. He has used gunfire either as a warning or as a means of drawing attention to an arrival or special occasion. It was not until the discovery of electricity and the invention of the

7

electric telegraph that he was able to begin to communicate over more than a few miles and to bridge continents and even the world. Even more important was his ability to communicate over such distances immediately and to get a relatively swift response, compared with the use of human messengers or the mail. As late as the 1950s, telegrams continued to be relied upon to obtain information about stock and commodity prices in overseas markets. (London, astride the Greenwich meridian, was a major center for relaying international telegrams as it bridges the time difference between markets in the eastern and western hemispheres.) Even with the relatively slow speeds of the telegraph services and the frequent need to deliver and collect the replies by messenger, commercial transactions could often be completed by brokers and other dealers while clients waited on their premises.

It was not until the invention of the telephone by Alexander Graham Bell, however, that man was able to communicate over great distances using the spoken word and receive a virtually instant response and reaction to what he was saying. Until then all forms of communication had involved the use of some form of code, which had to be agreed upon in advance. (It is sometimes forgotten, however, that the telegraph and other message services could be, and still are, used when language differences make direct speech communication impossible, and may sometimes avoid misunderstanding and lack of certainty.) Telephony since its inception has involved the conversion of sound waves to equivalent variations in an electrical current, but unlike telegraph services no prior coding is necessary before transmission. After many years communications technology has begun to turn full circle, as will be explained in a later chapter. The latest systems of voice communication again involve coding the spoken word, and again use light as a communications medium.

In the more than one hundred years since Bell developed it in the 1870s, the telephone has become the primary means of instant communication. Telephone systems have grown rapidly in size and complexity and the use of the telephone has become commonplace, spanning the world. In the United Kingdom four out of five households and virtually every business have the telephone and can call some 500 million telephones throughout the world by dialing or keying. Around 50 million calls are made on an average day involving countless billions of electrical connections and circuits. This service has come to be accepted by the public at large as commonplace, without regard to either its technical complexity or its logistical scale. It has become such an essential part of everyday life, and users' expectations have been so raised, that any failings or weaknesses are severely criticized. The dominance of "plain old-fashioned telephony," however, in the last decade or so has begun to diminish as other forms of instant communication have begun to be developed. Nonvoice communications have become at least as important to industry and commerce as telephony, and in a relatively short time will become even greater in importance. Businesses already rely heavily on telecommunications to carry vast quantities of control and accounting data quickly and securely, and are using electronic message networks increasingly as a quicker and more secure alternative to postal services.

The pace of technological change is accelerating with communications and computer technologies converging rapidly. New systems are being developed and overtaken by even more advanced ones almost before they can be put into practical use.

At the same time in the Western world users are increasingly being given wide freedom to decide for themselves the telecommunications systems they consider best meet their needs and the terminals most suitable for their operations. Increasingly Governments and telecommunications administrations are relinquishing their power to decide for users which system they should use and their power to restrict the choice of telephones and other terminals. For decades users have had little choice about their communications systems, having to take whatever the administration of the local telecommunications system decided was best for them. To some extent this was inescapable as there are always two ends to a telephone call, and communication is impossible if the two terminals involved are technically or operationally incompatible. Standardization was imperative for efficient operation, and where this was not ensured (as was the case for many years in international communications), a satisfactory service could seldom be provided. Telex services in different parts of the world, for example, operated at different transmission speeds. Complex interface equipment had to be used to store and retransmit the transmission from the faster working teleprinter at the speed at which the slower could accept. Failure on the part of the operator at the faster transmitting terminal to remember this could result at best in pileups of stored messages and long delays while the equipment cleared itself, and at worst to a complete breakdown in communications. For decades in the United Kingdom the development of a facsimile service allowing the instant reproduction of a document at a distant terminal was prevented by failure to agree upon common operating standards. Attempts are made internationally to agree on technical and operating standards for telecommunications, but any agreement tends to have only the force of a recommendation. Countries are understandably reluctant to encourage or insist on the adoption of standards while their home based telecommunications manufacturers are marketing systems overseas that do not conform to those standards.

Today users have increasing freedom of choice, but that means they now have to know far more about the potential of the technology and how best to apply it. They must also expect some limitations to continue to be placed on that freedom in the interests of technical and operational standardization, without which national and international communications are jeopardized. The relationship between users and the telecommunications administrations is changing, and this too will be the subject of a later chapter.

3

TECHNOLOGICAL REVOLUTION

Since the invention of the telephone by Alexander Graham Bell, the immediate and rapid spread of the availability of telecommunications, ultimately throughout the world, has had a dramatic impact on industry and commerce. The ability of firms to compete in world markets and their commercial viability have come to depend critically on management being able to obtain information immediately about operations, about the availability of raw materials and other essential resources, and about the demand for the firm's goods and services. Commercial success or failure is critically dependent on knowledge of what is happening and less on subjective anticipation and judgment of what might be. Firms have become dependent on telecommunications to obtain such information. Operations are often dispersed over a wide area or even worldwide, markets are seldom confined to the immediate neighborhood of the firm, and trading conditions often are affected by external factors about which the management must be constantly informed. Postal and telegraphic services, even for local communication, are not fast enough when the information is needed for day-to-day decision making, or to respond to sudden and major changes in external political, industrial, or commercial conditions.

In the last decade or so the technology developed for computers, the techniques of controlling the operation of electronic systems by internal programming, and the progressive miniaturization and very large-scale integration of a multitude of electronic circuits have been applied to telecommunications. This is rivalling and at the same time complementing the computer technological revolution itself. The two technologies have converged and will advance together. Terminals are becoming widely available, providing not only telephone facilities but also data processing on- and off-line, and electronic message services.

Systems are being developed to capture, store, process, carry, and retrieve information faster than they can be brought into practical use or than the very real and tangible benefits can be realized. At best this represents a wastage of scarce innovative resources; at worst it is a failure to harness the advancing technology to raise the quality of life. There are two reasons for this failure. Firstly there is a shortage of trained men and women able to plan and implement in detail the application of the new technology, as distinct from the innovators and system developers which computer science and electronics graduates usually aspire to become. Secondly, and of even greater importance, is the magnitude of the task facing the senior management and boards of firms in applying the new technology. This is no longer merely a matter of automating a manual process, for example replacing an internal manual telephone system with an automatic private exchange.

It often means changing the structure, organization, management, and sometimes even the nature of the goods or services produced by the firm. For example it is possible, by means of an electronic switching system provided primarily for internal telephone calls, to capture at some central point a mass of detailed information about production or field operations. This can lead to a centralization of management and erosion of the authority of line management. Alternatively it can lead to more responsibility and discretion being given to line management, in the confidence that if it is exercised wrongly it will become immediately apparent and can be corrected. Clearly a balance must be struck, but senior management must recognize the choice it has to exercise, and consciously decide how its management style should be changed to take account of the impact of the new technology. Investment in a new telecommunications system is no longer merely a matter of automating the connection of telephone calls. New systems have potential for raising the efficiency and productivity of the firm, but cannot be introduced without detailed consideration of their indirect effects on the working of the firm.

When introducing new systems the effect on personnel has to be considered most carefully and prepared for well in advance. The retraining, redeployment, and, if necessary, reduction in the number of staff employed must be planned well ahead and discussed not only with their representatives, but with the individual employees. Planned sufficiently far in advance and openly discussed with the staff, ways of resolving potential problems can usually be found and new systems introduced without jeopardizing good industrial relations. New systems often raise the quality of work, eliminating repetitious and boring tasks, and possibly making few physical demands. Apprehension about learning new skills and techniques or even changing jobs is normally discovered to be ill-founded, even by the middle-aged fearing they might not be able to cope. Reductions in staff numbers can often be met by normal attrition, adjusting the level of intake of new staff, the voluntary redundancy system in the United Kingdom, or early retirement. The specification, procurement, and commissioning of a new system requires a long lead time and there should be every opportunity to prepare for its acceptance by the staff affected.

Without the application of new technology the expansion in telecommunications would not have been possible. If automatic telephony had not been introduced half the population of a country today would be setting up telephone calls for the other half. In countless instances it is only by applying new technology and by employing existing staff more productively that businesses have been able to expand and grow. If, for example, banks had to continue to rely on the capture, storage, retrieval and movement of information about their clients' accounts on manual processes and paper records, commerce would have ground to a halt long since. Merely employing additional staff would not have resulted in the vast expansion in banking business and the speed of transactions required for it.

Telecommunications systems can now be very complex and represent a major investment and ongoing expense. The rapid development of the technology makes the timing of investment decisions particularly difficult, with systems commonly becoming obsolescent well before the end of a normal service life. Senior management often has to make the difficult choice whether to invest in a new telecommunications system now and take early advantage of the facilities available, or to postpone investment and await a more advanced system known to be under development. It often has to decide whether to buy a system outright, recognizing that it may be some years before it is able to fully exploit its potential, or alternatively to lease a new system which it may be able to replace in a relatively short period, paying of course in front loaded, heavy initial charges for the option. Because of the failure of new systems to immediately fulfill their promise and the time necessary to eliminate unforeseen teething troubles, major users have become understandably cautious about investing in them.

4

THE INTERFACE
BETWEEN THE
USER AND THE
SYSTEM

A telecommunications system, like any piece of machinery, requires some form of interface between the user and the system. The user must be provided with some means of communicating with the system; he must have some means of controlling it and of monitoring its operation so that he may, if necessary, give other instructions. There must be some form of input and output between the user and the system.

In a simple telephone system the input is processed by means of a microphone to transform the speech waves of the speaker into electrical signals, and the output is processed by means of an earphone to transform the electrical signals back into sound waves for the listener. The control of the system is by means of a switch, usually called the "switchhook," built into the hook or rest on which the earphone or handset rests when the telephone is not in use. In most elementary systems a hand generator is also provided to signal that a call is required or is ended. Finally, a bell is provided for the system to inform the user that a call is waiting for him. Such a simple telephone system can be constructed to be very robust and reliable. It will continue to function in hazardous conditions, and is still relied on today for emergency use in coal mines, petrochemical plants, and similar operations.

The microphone and the earphone, albeit with some technical development over the years to reproduce the speaker's voice more faithfully, have remained at the heart of the most advanced telephone systems. Initially the means of controlling the system was by turning the handle of the generator to signal the operator for attention, to ring a bell at the distant telephone, and to signal the end of the call. With the growth in the availability of the telephone and its use, methods were developed to set up and clear connections more quickly. For the last sixty years or so, systems have been controlled by means of the switchhook, and a dial to select a chain of connections to the distant telephone. The switchhook remains in some form, but the dial is fast being replaced by a set of push buttons as a means of selection.

Formerly, at the start of a call an operator spoke to the user and subsequently reported the progress of the call, or even that it could not be established. With the change to a completely automatic system, other means had to be developed for the system to communicate with the user. Various tones are used to indicate progress made in setting up a connection or to request some response from the user. A dial tone is used at the commencement of a call to signify that the system is ready to accept instructions, and a ringing tone to signify that the distant telephone is being rung. (It is not true, as many users assume, that they are hearing the required telephone ringing when they hear the ringing tone. The ringing tone merely indicates that the exchange is sending out a calling signal: The distant bell may not be actually ringing— the lines may be down or some other fault may exist outside the exchange.) An engaged tone or busy signal is used to indicate that the required number is in use or that there is no free chain of connections to the required number. In some systems the same tone is still used to indicate that the required telephone number or all the equipment is in use. (That the same tone is used in both circumstances is not always understood by users and has sometimes led to domestic disharmony.) In some instances, instead of simple tones, recorded or simulated voice announcements are used, for example to state that all lines to a certain town are in use. In the more advanced systems the communication with the user may be visual, in the simplest form by using an indicator light, in the more complex by displaying a message on a screen or other visual display. For example, many telecommunications companies use public pay phones which show the user how much of the money he initially deposited remains to be used. Additionally, telephones are becoming available which display the number of the telephone to which a connection is being set up, and once the call is established, display the time which has elapsed since it started—and for which the user will be charged.

The interface between the telegraph system and its users was initially some kind of key to send with, and some form of visual or sound signaling-device to receive the message. The earliest telegraph systems used a needle moving across a dial for signaling, but this was replaced very soon by a "sounder," in which a heavy brass arm was moved between metal stops to signal the Morse code in clicks. Sound signals could be read from a distance, unlike the visual display, and operating speeds were faster. The written message, however, had to be coded first before it could be transmitted, for example into Morse code, and then be decoded at the receiving end. Today the key has been replaced in telex and other message systems by a more complicated keyboard based on the typewriter keyboard, and the visual or sound signaling-system has been replaced by either a printed output of the message or its display on a screen. The coding and decoding of the transmitted message is done by the system and not the user. Attempts to replace the keyboard by some form of tablet on which the user could write his message and avoid the need for touch-typing skills have proved to have very limited practical application.

In the future, communication between the user and the system using a visual display may go further than merely displaying a message: Visual images of the user may be transmitted. It is not self-evident, however, that this adds to the effectiveness of communication. There is some evidence that seeing the person at the distant end inhibits, rather than enhances, the interchange of information. It is common experience that people will often say things over the telephone which they would not say in a face-to-face situation, with the inhibition of eye contact.

However, visual communication clearly does have worthwhile applications in such uses as the remote observation of road conditions or property. Today telecommunications are commonly used as a means of keeping property under surveillance to detect intruders, fire, or some other threatening situation.

The interface in this case is by means of some form of sensing device. Sensing devices are also being used to keep people under surveillance, for example the elderly and infirm living alone. The use of telecommunications for remote medical care using some form of sensing device is clearly a likely development, merely extending the techniques already used in the intensive-care departments of hospitals to patients under care at home.

The interface between users and telecommunications systems will continue to develop rapidly. Systems are being developed which respond to the user's spoken instructions, avoiding the use of dials, keys, and other controls requiring manual dexterity. The tendency will continue to be to make the use of systems "user friendly" and easier to use, and more important, to stimulate greater usage. No more than an indication has been given above of the trend and its possibilities. No matter how advanced the design of a system, in the last analysis its operational value depends on the interface with the user and the degree of skill, dexterity, and training he or she needs to use it efficiently. Whether or not to invest in a telecommunications system is likely to depend, in the last analysis, on the ability of intended users to cope with it and their willingness to use the interface. If the use of the system requires a high level of dexterity or precision, users will tend to make mistakes and through impatience and frustration, not use it. Many telecommunications systems have been designed for the business executive with a multitude of sophisticated facilities. The executive may have neither the time nor the patience to learn how to use the system, preferring to have a secretary cope with it. The slow growth in the use of videotext in the United Kingdom is attributable to the complex, albeit logical, procedure needed to obtain required information.

5

DATA
COMMUNICATIONS

Attention so far has focused on two forms of communication: Voice, in which the spoken word is carried by the system; and message services, in which the written or printed word is carried. In most developed countries the growth in the availability and use of the telephone in recent decades has been explosive, and for a time many European administrations were unable to meet public demand. The use of telecommunications for person-to-person and other forms of voice communication will continue to increase, but in most Western countries, at least, the very high rates of growth have waned.

Telex and other message services, particularly "electronic mail" with printed messages that are carried electronically and displayed at the receiving terminal on a screen, will continue to grow rapidly. Visual services, particularly systems for conferences between remotely situated groups of people, will continue to be developed, and possibly with time a significant demand will develop for telephones with a screen on which the other person can be seen.

For the foreseeable future, however, the dominant growth in the use of telecommunications will be in data communications. When computers were first used for commercial applications the data which they were designed to store and process was entered manually, transported to the computer center, and then fed into the machine by keyboards or other mechanical means. It quickly became apparent that if the power of the computer was to be fully exploited, remote terminals would need to be connected to the computer by the telecommunications system so that the data could be carried electronically from the point of capture. This telecommunications system could also be used to send data back to the terminals from the computer. For example, "hours worked" and other information could be sent into a central computer handling the payroll of a large organization from outlying divisions for processing. The computer in turn could send back at the end of the week or month individual pay checks for the local payroll office to use to pay the employees.

Data communications used to carry large volumes of data at high speed between computers, has already grown fast from such beginnings and in time will rival or even overtake the use of telecommunications for voice services. Perhaps of even greater importance, however, has been the growth of a wide range of electronic terminals to capture data and to transmit it over the telecommunications system to the central processor, and to receive data for local application back from the central computer. The applications are almost limitless, ranging from the capture of data at the point of sale, to complex industrial controls, and include a variety of domestic applications.

Data terminals at the point of sale in stores, for example, can be used not only to check the credit worthiness of a customer with a credit card company, but also to debit the customer's account directly. Information about the goods being sold can at the same time be sent to a central data bank where records of the stock held at the store can be constantly updated. If minimum stock levels have been reached, replenishment orders can be initiated over direct data links to warehouses or even suppliers. The financial savings from being able to operate with lower distributed stock levels can quickly offset the costs of such a data network. Financial information can simultaneously be sent directly to a central accounting system so that the financial performance of the store can be kept under constant review and departures from the budget can be brought to notice promptly.

The production of goods often depends on the manufacture of parts or subassemblies at a number of separate plants, the output of which have to be timed and coordinated to ensure that all that is required for final assembly reaches the production line when required. Data networks are used to control the level of output of parts and subassemblies and effect their movement to the production line at the right time.

In a development which may have domestic applications, every major hotel now has a data network to detect an abnormal rise in temperature or smoke, to sound local alarms, to close fire-prevention doors, and to call out the emergency services in the event of fire.

Data communications have particular value when speed is essential, for example in verifying the credit worthiness of a customer before completing a sale, or when information may have a relatively short life, for example market prices.

II

TELECOMMUNICATIONS SYSTEMS

6

TELECOMMUNICATIONS TERMINALS

Most telecommunications systems are provided for the use of the public at large and the users are normally charged for connection to the system, and according to how they use it. Some systems are provided primarily for internal use, for example internal office systems, but are usually also connected to the public system. A minority are self-contained and are provided for the sole use of a closed group of users.

All telecommunications systems, even the most complex, consist of a number of separate elements. There are terminals to act as an interface between users and the system. The terminals are connected by means of local networks to some form of interconnecting switches. In the public systems and the larger private systems the switches are connected to each other by means of a main network. In this chapter the features and operating characteristics of terminals will be described. The main types of terminals are voice, message, and data.

VOICE TERMINALS

The telephone remains the basic terminal to a telecommunications system, providing the interface with the user for voice services. With automation it was rapidly developed to become not just a means of voice communication, but also a means of controlling the system and of receiving messages from the system itself. Telecommunications systems are becoming increasingly complex. To make them more "user friendly," terminals are being developed which are operationally less demanding. To provide additional facilities and to help the user software is now being incorporated in the telephone itself. In many cases the software in the telephone is performing functions previously carried out by the switch, and can be more easily programmed to meet the particular needs of the individual user.

In a small manual telephone system with calls being connected by an operator there was no need for users to know the number of the person to whom they wished to speak. They merely asked the operator for the person by name, relying on the operator to know which line would connect them. In the larger systems the operator could not remember all the lines to which calls should be connected and a system of numbering had to be introduced. The need for a numbering system was inevitable with the growing use of the telephone, but has long been a source of operational difficulty and has added considerably to operating costs. As the system grew in size users have often been reluctant to lose the option of asking the operator for the persons they wanted by name. The costs of recording and publishing directories listing numbers are substantial, particularly where numbers are constantly changing. (On average, between 20 and 25% of the entries in a British Telecommunications directory change between issues.)

The way numbers are shown in a telephone directory can be a cause of highly subjective complaints with little objective operational foundation. For example, an entry does not contain any more information about a person, or the person's address, than is essential for positive identification. To minimize printing costs, house or apartment names and postal codes are omitted, along with honors and decorations. Titled or degreed individuals often object to what they regard as a failure to present them in a prestigious way. Numbers which appear to have particular significance have been prized and sought after.

Failure of users to remember or to use the numbers correctly can inflate operating costs, and irritate and inconvenience other users. It is not surprising, therefore, that development of telephone software has focused on this aspect of telephone usage.

As well as having to remember the internal business numbers of people frequently called, with the growth in national services and their interconnection into worldwide networks the user is now required to identify the number he wants from literally millions. (In the United Kingdom, with international direct dialing when the user picks up the phone he has well in excess of 500 million telephone numbers which he can dial. Any one of those numbers may give him access to an internal telephone system and several hundreds more internal extension numbers.) He not only has to select and remember the number he needs, but he also has to dial (or now more commonly, key in) the required number correctly. The number of digits in the number may be large and the risk of error significant. For user convenience and also to minimize the incidence of wrong numbers, telephones have been developed with built-in memories. The user can store in advance those numbers which he calls most frequently, and subsequently obtain them by keying the short number codes he has personally assigned to them. Some terminals store the numbers of the persons regularly called, thereby eliminating the need for users even to remember short number codes. The call is set up by merely selecting the name of the required person and pressing a button.

With worldwide automatic dialing/keying, a wrongly selected number may engage a chain of connections to the other end of the country or possibly the world. Apart from using the system ineffectively and possibly denying its immediate use to others, if the number called answers, the user is charged. The telecommunications system cannot distinguish between calls established by users dialing/keying numbers incorrectly and those that are correct, and will charge the caller for the mistakes as well as for the correctly dialed calls. The penalties to the person who makes a limited number of calls are minimal, but for the large user the expense could be significant. Careless use of the numbering system by employees in a large organization can significantly inflate a firm's telephone bill.

Many telephones now incorporate a repeat call facility. The last number keyed is automatically stored. If it is busy on the first attempt, instead of keying the whole number again a single key can be pressed to set up a repeat attempt. Alternatively, if a follow-up call is required to the last number spoken to, it can also be set up with a single key stroke.

Telephones have been developed that incorporate a small visual display. While the telephone is not in use this can function as a digital clock, and during a call can show the time elapsing while the call is in progress. This provides the user with an indication of the charges he is incurring (in some versions the actual charges are displayed) because users are understandably concerned that they do not know while the call is being made how much it is costing. The software to make this possible may have to be programmed to compute the variables of the particular telephone; the distance between the two telephones; the time of day; and the day of the week or year.

Visual displays may also be used to show the user the digits he is keying as he is setting up the call, and thus bring to attention any miskeying. Displays may also convey limited information to the user either from a central data bank (possibly the number he requires in response to a directory enquiry) or a message from another caller, thus serving a purpose similar to answering/recording machines.

In recent years more attention has been given to the physical design and appearance of the basic telephone, and provision has been made for its use other than on the desk or table top. Telephones concealing the same internal hardware and software come in a wide ranging selection of plastic cases. The handset incorporating the microphone and the earphone remains the same, albeit with some variety of design that must remain within the limitations of having to bridge the distance between the ear and the mouth. Alternatively some telephones have a loudspeaker and a more sensitive microphone built into the case which can be used without the handset. These are particularly valuable when more than one person at the terminal wishes to hear or take part in the discussion. Possibly their main attraction is the freedom they give to write, look up files, and move around while conversing. There are of course limitations on their use, for example they cannot be used in noisy surroundings or when the user does not wish the conversation to be overheard. There remains on the part of some users a reluctance to speak when they know that such a "loudspeaking" telephone is being used, feeling that it in some way mars the intimacy of the conversation.

An answering machine may be attached to the terminal telephone to play a prerecorded message to callers while the user is away. Callers may be invited to leave a message on the machine. While these machines can be useful in the home they do indicate that the premises are probably unoccupied and could invite break-ins. For business use they have value as a means of taking orders after normal business hours. Used during normal business hours, for example to say that the user is temporarily out of his office, they can convey an impression of discourtesy and ineffective secretarial coverage of absences.

Telephones have been available for some time which need not be connected to the system by wires or cable. The first models were radio telephones intended for use in the car or for other mobile purposes. These have been developed to the point where they can provide facilities virtually equivalent to the basic conventional telephone. Two further developments of the concept are of particular importance: cellular radio and the cordless phone.

In some terrain where a radio-telephone equipped vehicle is in the "shadow" of a hill or large building, the radio path may be blocked and operation interfered with. The development of cellular radio has been of great significance therefore. This involves the provision of a large number of low powered transmitting/receiving radio stations linked by a central computer. The computer constantly monitors the location of all terminals, and communications with them are automatically maintained via whichever cellular radio station provides the best radio path. A small telephone is carried with the user wherever he goes and provides a fully mobile service, including some nonvoice services.

The costs of cellular radio terminals are at the moment comparatively high. As inevitably the price is brought down, however, a rapid growth in their use can be expected. Having an instant means of communication, being instantly able to be contacted wherever the user goes, and not having to go to a terminal physically tied to the end of the telecommunications network are obvious advantages. Cellular radio terminals are already being developed for a variety of nonvoice services. For example they are a means of keeping persons and property under constant surveillance wherever they are.

Linkage by means of short-distance low powered radio between the end of the wired or cabled system and the telephone is the basis for the cordless phone. The terminal telephone can thus be carried from room to room or for short distances outside the premises. The continued miniaturization of components and reduction in the size and weight of the telephone, making it truly transportable, are necessary to develop fully the potential of the cordless telephone.

MESSAGE TERMINALS

Another group of terminals provide message-type services—services carrying the printed or written word. The earliest form of such services was telegraphy, using the Morse code and a telegraph key or some other form of code and keyset. This was slow and required a marked degree of skill by the operator, but it could be used for communication using networks or radio paths of very low quality. Morse code was used for international cable services and for ship-to-shore communications until comparatively recently. To improve the speed of transmission and to take advantage of the development of networks and radio channels with a much improved transmission quality, Morse was replaced in public telegraph services by teleprinters using conventional typewriter keyboards. Such services could be used for private point-to-point communications and required little more operating skill than that of a typist.

Telex, a development of point-to-point teleprinter services into a public switched service, has been available for decades. Connections can be set up as required between distant terminals allowing messages typed at one terminal to be received in print at the other. Initially the service was provided over the telephone network: A telephone call was established between the two terminals and then both were switched over to teleprinter. The signals from the teleprinters were converted to tones in the speech frequency range before being sent over the network. At the receiving terminal the incoming tones were converted back to direct current pulses to operate the teleprinter. The calls always had to be set up via the operator and special arrangements had to be made to protect the connection from interruption, and to safeguard the quality of transmitted signals.

For some time now telex services have been operated on specially provided networks and via switches used only for the connection of telex calls, enabling a better quality of service to be provided. The design of the terminals and their performance have been further improved by the application of electronics technology and software programming. From the user's point of view a disadvantage of the telex service is still the need for touch-typing skills if the service is to be used efficiently.

Facsimile is an alternative which has been available for some time. In this service a document, either prepared by hand or typed, is scanned at the outgoing terminal, carried electronically by the telecommunications system, and then reproduced at the distant end. The quality of the received copy was not good until recently as there were no agreed upon technical standards, and with different makes of terminals at each end the quality was rarely satisfactory. Again the design of the terminals and their performance has been improved significantly, and operating standards have been agreed upon.

Both telex and facsimile were used primarily for communications within the same organization, in contrast to telephony which almost from its inception has been a truly common user service. In the last decade or so there has been a growing trend to use message services for communications with other organizations in the same way as telephony.

Word processing, allied with electronic mail, is the latest message type service. It started with the development of ways to type, amend, and correct text on a screen before its production on a high speed printer. The text is stored in a data bank, can be recalled if needed, and can be printed with variations, for example to give a personal connotation to a circular letter. Given the storage of such text electronically in a data bank, it was a logical development to carry it electronically over the telecommunications system to a second data bank, where it could be retrieved and be printed if needed. With appropriate operating protocols, such arrangements have developed into electronic "mail" and could rival, if not replace, telex as the primary means of sending messages. Like telex it has the disadvantage of needing touch-typing skills for fully efficient utilization. As with facsimile terminals, connections between word processing terminals can be set up over the public telephone network, the terminals transmitting tones in the voice frequency range for conversion back at the receiving end into direct current pulses.

The speed with which messages can be sent direct to the intended recipient by electronic mail, and answers expected back, appears to be having some unforeseen effects on management behavior. Normal mail services allow some hours for replies to be considered and prepared between receipt of the morning post and the last mail pick-up at the end of the working day. With electronic mail messages can be sent and received virtually immediately at any time during the day and replies expected within hours or even immediately. The "thinking time" of managers may be eroded and replies have to be given without adequate time for reflection. There also appears to be a tendency for messages—sometimes not well prepared—to be originated by electronic mail which otherwise would not be typed and sent by post. This electronic equivalent of "junk mail" is already leading to electronic mail addresses only being revealed to selected colleagues.

Electronic mail appears unlikely to replace telex in the foreseeable future. As the name implies, it is more likely to replace communication of essentially personal messages by surface mail. Telex will continue to be used for the fast transmission of data-text prepared in advance by specially trained staff and often broadcast to a number of addressees. Telex has now established itself as a legally binding way of transacting business while the status of electronic mail has yet to be established in that regard.

DATA TERMINALS

Telecommunications systems have been used since early in the development of computers as a means of carrying data captured at a remote point to a central data bank, and for carrying data back to the initiating point. Initially all data were coded numerically before being sent to the computer and a simple keyboard was used as the terminal. Demand soon developed for a more complex form of data input and output, and full alphanumeric keyboards and printers were introduced, with additional keys for specific operations. Essentially data terminals convert the manually prepared data into electrical signals which can then be carried to the central system.

Increasingly such data terminals are incorporating minicomputers with software which can be locally programmed. Data can be stored, processed, retrieved, and the terminals can be operated independently to a growing extent. The use of the telecommunications system to carry large volumes of data which is not required quickly is tending to diminish. It is often less expensive to carry the magnetically-stored data by courier, especially over distances which can be covered within the time by which the data is required.

With the growing commercial reliance on these information systems, companies have become increasingly concerned about the security of their stored information. Extreme measures are taken internally to protect data against accidental loss, mutilation, or criminal activities. Understandably, companies are reluctant to transmit highly sensitive data over telecommunications systems, fearing that at best errors may be introduced during transmission, and that at worst the data may be accessed by individuals with criminal intentions. There is a tendency to move nontime-critical, highly sensitive data between terminals by courier as well as bulk nontime-critical data.

The telecommunications system tends to be used for immediate access to data stored on remote large central data banks which it is not economical to copy and hold locally. It is better to store data which is subject to frequent updating by this method, on a central data bank to which terminals can have immediate on-line access, rather than use the more expensive alternative of constantly updating the data held at each terminal. Telecommunications companies have offered special low rates for the use of their systems at night in an attempt to retain the transmission of large-volume, nontime-critical data and at the same time to take up system capacity which would otherwise be unused.

The following example of the use of both the physical shipment of data and the instant electronic transmission of data comes from the banking field. Many of the transactions in the branch of a bank need not be processed in the central customer accounts until the end of the day. They can be stored at the branch and the tapes, or other magnetic forms of storage, taken by courier to the central accounts department after the end of banking hours. If, however, a customer wants to know the state of his account, the bank's central data bank is accessed over the telecommunications system and immediately can be used by the staff at the local branch. Cash dispensing machines are another example of using the telecommunications system for short duration, immediate data communications.

The fastest growing use of the telecommunications system for short duration data transmission is in such fields as airline reservations, with the booking clerk checking the central seat-reservations data bank and making an immediate reservation and printing out the ticket while the prospective passenger is waiting.

Other forms of data terminals like the cash dispensing machines still use a simple numeric keypad with the addition of a few specifically designated function keys. In some cases the keypad of the terminal telephone itself is used for the input of data. In other cases a special data-reading device is incorporated into the terminal and the need for keying eliminated. Telephones have been developed with a magnetic strip reader for instant verification of credit cards. A sales assistant presented with a credit card by a customer makes a normal telephone call to the credit card center. Having established the connection, the customer's credit card is passed through a slot at the back of the telephone and the data stored magnetically on the strip on the reverse of the card is automatically sent to the credit card center for validation. Advice on whether to allow the purchase or to impose some limitation on the use of the card is received back by the sales assistant either aurally, or displayed on a small visual display strip built into the telephone.

Petrol pumps have been developed for use in conjunction with credit cards: The validity of a card on insertion is first confirmed over a data link to a central computer. The delivery nozzle is then released, and the motorist takes whatever volume and grade of fuel required. The charges are registered at the pump in the normal way, but on replacing the nozzle they are transmitted and debited against the motorist's credit account at the central data bank before the credit card can be removed. Such electronic fund transactions, apart from improving the cash flow of petrol stations, are a significant counter to crime in avoiding cash transactions.

Some data terminals are built into other terminal equipment and the data are captured and transmitted to a central point as a secondary operation. For example, the conventional cash register at the point of sale has reached the point of development at which it is not merely used to total the customer's bill, including calculating the value of purchases by weight. It is now used to capture and transmit information instantly to the central accounting department for cash flow and control purposes. It is also used to collect data about the movement of goods, providing instant stock records. Such data is carried by a local network within the store for accumulation, and later physically transferred to a main computer serving a number of stores. Alternatively, depending on the relative cost benefits, it is carried electronically over a more extensive local network to a central computer data bank.

Telecommunications are being increasingly used to link stand-alone personal computers to business systems, or to access common user data banks from them. Many of the computers and systems have the necessary interfaces built in. Such facilities are now being used not only for the transmission of user data, but also to load the stand-alone computers with software programs held at a central point.

OTHER TERMINALS

There are a range of terminals available other than the voice, message, and data terminals discussed, and the number of devices and applications is constantly growing. For some time systems have been used to keep properties under surveillance. As well as sounding an audible alarm on the premises when triggered by intruders, they are used to transmit a signal over the telecommunications system to alert a central security organization. Some automatically make a call to the police's emergency phone number and on being answered transmit a prerecorded message, stating that the premises have been broken into and requesting assistance. Others are connected to a security organization which constantly monitors the state of the alarm device and can be alerted more quickly and with more detailed information about the intrusion. Surveillance systems are of course only effective if the telecommunications system itself is secure. The system itself must also be protected to prevent the alarms being made inoperative or from detection that they are no longer providing surveillance. The more advanced systems provide continuous visual surveillance from a central point, and are usually more easily provided where the premises are served by a broad-band coaxial cable system.

Other devices are becoming available for a range of telemetry and remote control purposes, for example constantly measuring the level of water in reservoirs or liquids in storage tanks. Other devices have been developed for the remote measurement and control of electricity: Electricity-supply authorities need to be able to control or shed load if demand exceeds generating capacity, and they are concerned with being able to vary charges with more precision to encourage the shift of consumption to periods when there is an excess of generation. Remote telemetry offers this ability, which the existing metering arrangements cannot.

The provision of broad-band coaxial and other systems for cable television is likely to lead to the development of these and other services as part of the package offered to attract subscribers. One of the most exciting possibilities is the use of such systems for home study programs. The student is able to study at home, communicating over a data link with a central computer which checks his or her responses to questions and only allows access to further lessons when it is certain the course has been understood up to that point. The cable television companies faced with heavy investment could well seek ways of raising the utilization of the cables and reducing unit costs by adding these services.

The use of interactive tele-text services offering the user access to central information data banks is likely to grow quickly. To date, however, the use of the residential telephone to gain access to a central data bank from which information could be selected and displayed on the home television has not developed as quickly as was expected. There are a number of reasons for this. Inadequate thought was given initially to the presentation of the information on the screen, or the need to present it differently from its presentation in print. The selection of the particular frame containing the information required by the user often required a systematic and logical approach which the lay user could not or would not follow. Much of the information available was not time critical and could be obtained from other sources without incurring the relatively heavy telephone and access charges.

Whether these obstacles to the development of common user tele-text services would have been overcome by offering interactive services such as home banking and home shopping remains to be seen. Technically such systems are sound and the actual manipulation of the terminals is straightforward. Their growth potential for the foreseeable future probably lies with providing more specialized information to users better able to cope with information retrieval protocols, and presenting information that meets their users' special needs. The services that provide commodity and stock market prices and similar financial information have developed faster, particularly for closed groups of users who have sole access to the information.

Potential users have also probably been confused by a multiplicity of ill-defined terms, e.g., viewdata, videotext, teletext, etc. The British television broadcast services offer viewers access to information carried over the air in that part of the transmitted bandwidth not required for the broadcast sound and vision program. It is essentially a broadcast service with a volume of information being constantly transmitted and centrally updated. The user can decide which particular frame of information she requires and when that particular frame is received by her receiver it is stored and displayed continually on the screen. If the viewer wishes to refer to another frame she must direct her receiver to release the current frame and replace it with the required frame as it is broadcast, and store that instead. The selection and display of information is slow, but the user is not charged for accessing the information.

In contrast a more selective and quicker access is provided over the telecommunications network, again using the broadcast television receiver as a terminal. The user gains access over his normal telephone line to a number of central data banks, from which he can select the information he requires. He is also often able to send data back for central processing. For example, having checked which airline has seats available to a particular destination, he may be able to make a reservation with the airline there and then. In addition to the charge for use of the telecommunications network, he may have to pay a subscription charge for the service, and in some instances he may be charged every time he accesses certain information.

A third broad category of publicly accessible data bases is provided over telecommunications networks, but access is restricted to closed groups of users. They are normally provided with more advanced terminals than the commercial television receiver, and the use of such services requires specialized knowledge and operating skills. Information about money and commodity markets is often disseminated over such networks. The providers of such services, whether the telecommunications networks themselves or those renting restricted networks from the telecommunications operators, often charge a high premium for the information being made available that reflects its commercial importance.

Telecommunications companies in the United Kingdom are now being compelled to rent their systems to other operators, who may then use them to market services to third parties. The significance of this will be discussed in a later chapter, but independent tele-text services are an obvious example of such possible value added services or leasing arrangements.

TERMINALS–SUMMARY

The development of terminals for telecommunications systems is growing fast, partly as the technology advances, partly as the regulation of terminals is relaxed and is thrown open to competition, and partly in response to demand. No more than an indication has been given of the broad categories of terminals available. As this book is being written undoubtedly many more are becoming available. No one element of a telecommunications system can be regarded as more important than any other. Efficient operation depends on the technical and operational integration of the entire system. In the last analysis, however, whether the user is able to take full advantage of telecommunications depends intrinsically on terminal design.

At the moment terminal technology is ahead of its application, and possibly too little thought is being given to what users need terminals for and how they will use them. Some terminals are in advance of users perceiving a need or having an awareness that that need could be met. Others have been developed with no real need in view and although technologically impressive are unlikely to continue in use for long after an initial honeymoon period with the user. Others require a degree of understanding and operational competence which users have not yet acquired.

Visions have been created of a society in which man can meet all of his needs through some remote terminal, working, shopping, and entertaining himself without leaving his home. Unless man ceases to be a social, gregarious being, it is doubtful if such will ever come to pass. Businessmen still find it necessary to travel long distances and to meet face to face, rather than use the remote visual conferencing facilities already available. Whatever may be said about the irritations of shopping it is questionable if supermarkets will give way to remote shopping via some tele-text system. Much is possible, but a down-to-earth sense of realism about the nature of man and his need to meet other human beings face-to-face, and to have direct contact with others, is essential if much of the technological development of terminals is not to prove unproductive.

7

LOCAL TELECOMMUNICATIONS NETWORKS

INTERNAL NETWORKS

Early in the development of telephony the demand arose for some means of using the system at more than one place in the home or office, and wiring was installed to make more than one telephone available. This arrangement continues today, but often the internal wiring terminates on sockets and the telephone cords terminate with plugs so that they can be moved from one position to another with the number of telephones needed accordingly reduced.

Not surprisingly, this arrangement of multiple phones was quickly developed to enable users to speak internally from one point to another, with different methods of establishing the connection. In the most simple arrangement a push button or key is provided to ring a bell at the second terminal. In more complex arrangements, incoming calls are answered on one "main" telephone and are only connected to the second if appropriate. Such an arrangement clearly can be used to ensure that calls are filtered by a secretary before being connected. In the reverse direction the secretary can obtain outgoing calls and recall the principal user when the distant person is on the line. (Such an arrangement often results in competition between secretaries to avoid having the "boss" on the line first, in a form of status gamesmanship that wastes time and inflates telephone costs.)

Other arrangements provide two outgoing lines from the same telephone. Calls can be held on one, while the user switches to a second to make an enquiry before coming back to the first line. A multiplicity of special user needs led to the development of numerous complicated arrangements. Some were satisfactory but many had a high fault liability and were often a source of noise and other operational difficulties.

At an early stage some means of answering calls coming in from the public network and connecting them to the appropriate internal "extension" telephone was needed. With only a limited number of internal telephones this could be done with some form of manually operated switches at one of the telephones specially equipped for the purpose. Larger installations needed a "private branch exchange" with an operator to complete the connection with plugs and cords.

Arrangements also have to be made not only to connect incoming calls from the public network to the appropriate extensions, but also to connect extensions to the public network when outgoing calls are required. With short distances between the telephones and the point at which incoming calls are answered, the costs of wiring or cabling every telephone back to a central terminal or cross-connecting "frame" were and are acceptable only in relation to the extent to which they are used. The costs of connecting every internal extension to the public exchange can seldom be justified, and certainly not for any but modest-sized installations. The main telephone or private branch exchange is thus used not only to filter and direct incoming calls from the public network to the appropriate extensions, but also to connect the extensions to one of a limited number of outgoing lines to the public network.

Simple telephone cabling using twisted copper pairs is normally adequate for voice and most nonvoice services, and the cost of coaxial or optical fiber for internal networks can seldom be justified. If for a particular purpose a larger bandwidth/higher transmission speed is needed, special cabling is usually provided. The local network so far described is what might be termed a star distribution system with each terminal being individually wired back to a central point, albeit with larger cables being used as the wiring is brought back to the cross-connecting frame. An alternative is to use a ring-type distribution in which all terminals are connected to a common cable. This requires far more complex terminals and certainly for the foreseeable future is only justified for higher performance data communications. In some instances an internal network may be extended to terminals outside the premises, renting lines in the public network for the purpose.

LOCAL NETWORKS TRANSMISSION

In any telecommunications system it must not only be possible to establish contact between terminals, but when the connection is established the quality of the received speech or data must be intelligible. The quality of the transmission must be to a minimum standard, but equally important, costs should not be incurred in achieving standards of transmission higher than is commercially necessary. Reference was made above to the use of the main telephone or private branch exchange to concentrate outgoing calls on to a limited number of lines into the public network to reduce costs. In the more extensive local networks arrangements are made not only to safeguard transmission standards, but also to carry more than one connection over a single path simultaneously and thereby reduce local network costs.

Sound waves are converted by the conventional telephone terminal into an electrical signal which varies in frequency and amplitude, within the limitations of the system reproducing the original voice or other sound source. The earphone at the distant telephone converts the incoming electrical signal back into a sound wave, again as faithfully as possible reproducing the original sound. Internal local networks connecting telephones relatively near to each other using twisted pairs of copper wires enclosed either individually or with others in a cable sheath, are quite capable of carrying such "analog" signals and of reproducing them at the incoming terminal undistorted and loud enough for fully effective communication.

When such signals are carried into the local external network or further afield, however, they deteriorate. The received sound waves become distorted, suffering a loss of clarity and volume. In various ways this is corrected; on the longer circuits, for example, copper wires are used with a bigger diameter, reducing the resistance of the conductor and therefore transmission loss. Cables are designed and engineered to minimize interference from other circuits, and if necessary amplifiers are used at intervals to increase the power of the electrical signals.

Over longer distances other means of ensuring the quality of the received sound have to be used. These involve converting the "analog" electrical signals from the telephone microphone into other forms or transmission modes, and then back again into analog electrical signals for final conversion to sound waves by the earphone at the receiving terminal. At the same time that this technique maintains the quality of voice signals, many more separate signals can be carried simultaneously compared with a pair of copper wires. More important, the speed with which data and other nonvoice communications can be carried is significantly raised.

Two other techniques are of particular significance. The first is the use of pulse code modulation in which voice and nonvoice communications are coded into on/off pulses before being carried and then are decoded at the receiving end. The second is the use of coaxial cables capable of carrying a broad band of frequencies. These can be used for very fast data communications; for visual services requiring a much wider band of frequencies than can be carried by a pair of copper wires; or for carrying a large number of simultaneous voice services. Both of these techniques may be applied at the same time and can have a major influence on the costs of communication, reducing unit costs in some instances, but in others incurring costs which are not justified by the limited use made of the higher performance network. These techniques will be discussed in greater detail in a later chapter on main networks (see Chapter 9).

EXTERNAL LOCAL NETWORKS

In the earliest days of telephony the only way of connecting terminals externally was by bare copper wires carried overhead on wooden poles. With the rapid growth of the system it soon became impossible in towns and cities to continue with such a network. Telephone cables were developed containing up to several thousand pairs of wires. These were usually laid in earthenware ducts under the streets, gradually branching out into smaller cables and ultimately terminating at distribution poles where overhead wires could be taken to as many as thirty or so terminals. This remains the most common form of external local network. Methods of construction have been improved over the years: The two copper wires are now normally contained in a single sheath and attempts have been made to make the poles less obtrusive environmentally. In town and city centers where the networks are more concentrated, branch cables may be taken into buildings entirely underground, or the cables are terminated on the walls of buildings and extended by leads run along the surface to convenient entry points.

The local external network is the most vulnerable part of a telecommunications system. It is in a hostile environment in terms of both the weather and deliberate or accidental damage and interference. It is constantly being extended and altered as terminals are moved or added, and has a relatively high fault rate with at the same time a relatively low speed of repair. The growth of cable television has led to the use of coaxial cable, significantly less vulnerable, and in time offering an alternative local network for connecting nonvoice and possibly voice terminals. The final connection into the building, however, is likely to continue to be the weakest link in the chain, and thus another reason for the importance of the development of cellular radio.

Local networks represent possibly 30% of the capital costs of telecommunications systems, and their continued maintenance is a major element of the ongoing costs of operation. Rates are levied by local authorities for running cables and siting other external equipment on public highways and property. Telecommunications companies often encounter difficulty obtaining permission—"wayleaves"—to place their equipment and external cables in the most economic and best operational position. This equipment on average is only earning revenue five minutes in every twenty-four hours. Possibly 20% or more spare equipment and cable has to be provided to cater for growth and the constant movement of terminals. Costs are high for engineering to minimize damage due to the weather, vandalism, and accidental damage, and repair costs are also high. Ways are constantly sought to reduce capital and recurrent expenditures on local networks: The potential commercial benefits are substantial, but the technological and operational problems to be resolved are formidable.

8

TELECOMMUNICATIONS SWITCHING

In every telecommunications system some means must be provided for interconnecting terminals which are not permanently connected together, as needed and when required. This in all but the smallest installations is done by means of an exchange, either manually operated or automatic, now commonly termed a "switch." The exchange provided for the exclusive use of a particular firm has traditionally been called a "private branch exchange" to distinguish it from the public exchange provided for general use. Terminology varies from one administration to another, but the principle remains. The exchange or switch is not only a means of interconnection, it is also a means of raising the loading of the network and other equipment, the use of which is generally shared by terminals. This has a major bearing on the costs of systems and will be discussed in a later chapter.

METHODS OF SELECTION

In all but the smallest installations, terminals must be given an identifying number or "extension number." The number may be the number of the room in which it is located or be in a series associated with the floor of the building on which it is located. The numbering scheme for larger installations, if carefully designed, can help users remember and make fewer errors when selecting the terminals they wish to be connected with. A badly designed numbering scheme can be confusing to users and inflate the costs of the system.

There are essentially two methods by which required numbers are selected automatically. The first, based on the way in which operators on large switchboards selected the numbers to which connections were requested, is the "step-by-step" method. The operator on being asked for a number first identified the thousands group and moved his or her hand to that part of the jack field of the switchboard. The operator next identified the hundreds, again moving the hand with the connecting plug and cord to the one hundred jacks in which the required number was located, and finally moved up to the tens and along to the required number, before inserting the plug.

The step-by-step automatic systems now being superceded are based on a similar method. As the user dials or keys the required extension number a path through the exchange is established step by step, first to the thousands part of the exchange, then the hundreds, and finally to an electromagnetic switch which is stepped first to the tens and then to the required number.

Such systems are slow in operation, and communications are often noisy and of poor quality. As the public system is being modernized the limitations of the old step-by-step electromagnetic equipment are becoming more apparent. It is now possible to key as quickly as the required number can be thought of and the finger can be moved to the appropriate keys, but the time taken to establish the connection can be by comparison inordinately long. On some connections the required pulses at ten per second can be heard still being sent into the equipment long after keying is completed. On some calls the quality of the received speech is extremely good. (Often it is on the long distance calls going via satellite or other advanced systems that the quality of the transmission is best. In contrast calls over short distances can be very noisy, the equipment used having not yet been replaced. This is often most apparent on "phone-in" radio and television programs.)

The second method of selection is by means of a matrix consisting of a multitude of cross connection points. The digits dialed or keyed by the user, instead of directly operating a chain of electromagnetic switches, are used instead to identify the cross connections to be made in the matrix to establish a path to the required number. Connections are established quickly and communications are of a high quality. Switches are virtually entirely electronically based, and the transmission path has no moving parts from which so much of the noise in the old step-by-step systems comes.

The selection of the cross connections to be made in the matrix is controlled by software which can be programmed to provide a range of additional facilities for users. Switches are still being developed further and as this book is being written, existing equipment is becoming obsolescent after a very limited operational life. The same technology is used for both private branch exchanges and public exchanges, and the former can now be designed to provide many of the facilities which previously users could only obtain—and pay for—via the public exchange.

ACTIVE AND PASSIVE SWITCHING SYSTEMS

Step-by-step switching systems were designed to operate under the direct control of the user. Whatever the number dialed or keyed, a chain of connections is established to that number. At the same time, whatever the communication sent, this is reproduced at the distant terminal as faithfully as possible. The systems are thus essentially passive. The only exception to this was the introduction of equipment to make the numbering arrangements easier to cope with for the user in the systems serving large cities, and later for directly dialed longer distance calls. In the big cities the same alphabetical or numerical exchange code was dialed or keyed preceding the required number no matter where a call was being made from in town. Similarly on directly dialed longer distance calls the same numerical exchange code was dialed wherever the call was made from. The equipment then examined the code dialed or keyed and was wired to select the optimum path through the system to the distant exchange, replacing the code dialed or keyed with the appropriate digits to set up the path through the step-by-step network. This inevitably introduces a delay into the establishment of the connection, at times so long the caller hangs up, prematurely and incorrectly assuming that there has been a system failure.

Other refinements were in time added within the limitations of electromagnetic technology. For example arrangements can be made to divert calls for a particular number to another, or to the operator. This involves, however, making changes to the wiring of the system, and cannot be done at short notice. Doctors for example often need to have their incoming calls diverted to colleagues' attention in their absence. The arrangements made to provide this facility have never been entirely satisfactory, and are a constant source of operational difficulty. They are expensive to make, often need to be set up by a maintenance engineer, and the charges are accordingly comparatively high for the service rendered.

Message switching systems were refined to allow machines with different operating speeds to be connected together, or for errors introduced in the transmission of the message to be detected and corrected. Again these were never entirely satisfactory because in the worst cases they involved the use of mechanical devices, the reliability of which were never good. International message services were heavily in demand and the rate of growth high. The potential commercial benefits of immediate international telex communications quickly became apparent, but the frustrations of the poor service which could be provided with a step-by-step electromagnetic system were a constant source of complaint and embarrassment to the telecommunications administrations.

In recent decades refinements have continued to be made. For example, stored programs and software have been incorporated which take over the control of the call from the user. If for any reason the path through the network selected initially by the code dialed/keyed by the user is unavailable, an alternative will be immediately sought and the connection made without the caller being aware that this is happening. With some limitations, changes may be made to the program to divert calls to other numbers when required without altering the wiring of the system, and a better service provided for users wanting their calls diverted.

The convergence of telecommunications and computer technologies in the last decade or so has had a dramatic impact on switching systems. Selection of the required number by means of a matrix using electromagnetic switches was a limited advance from step-by-step systems. In a *fully* electronic exchange the matrix is contained in integrated circuits which can be operated at very high speeds and with very few moving parts. As well as making connections very quickly, the main cause of noisy communications is eliminated.

Such systems are active, registering the user's requirements, but setting up connections under the control of built-in programs. The programs can be changed under the instruction of the user to carry out a variety of tasks. The user for example may specifically instruct the system to transfer any call for him to another number where he will be located temporarily. The system may store numbers which are busy when first dialed and be instructed to make further attempts after a pause. The systems which have been developed offer more facilities than the user at present possibly recognizes as having potential benefit. Some of the facilities may in time prove to have limited value or to be making the system more complex than it need be. Other facilities are being incorporated into the terminal software and can thus be more personalized and better designed to meet the individual user's needs.

Modern systems used for nonvoice services can be programmed to convert the transmission of one terminal to a different form which will be acceptable to another and allow, for example, computers with different operating protocols to work with each other. The benefits have been particularly striking in international message services.

COMMON USER AND DEDICATED SWITCHING SYSTEMS

The switching system has two basic functions. It is, first, the means by which the user makes known the identity of the terminal with which communication is required and by which an appropriate chain of connections is established. It has a second and equally important function in that it is not known in advance when and between which terminals connections are required. The provision of sufficient equipment to ensure, however, that whenever a connection is required a free path through the system can always be found would be uneconomic. Only sufficient equipment is provided to ensure that at a specified level of usage not more than a certain percentage of calls shall fail due to all appropriate paths being in use. The second basic function of the switching system is thus to allocate the equipment available in such a manner as to ensure that the desired standard of service is maintained. In other words it allocates the equipment provided to achieve the optimum loading of the system. This is the essence of a "common user" network and in this way costs are minimized.

The alternative at the other extreme is to provide a terminal with a direct connection to another, and bypass the common user network with a "dedicated" or "private" circuit. Such circuits are likely to be unused for much of the time, and the benefits of assured instant availability are obtained at the cost of an uneconomic utilization rate. Commonly a compromise is provided, where a group of terminals is served by a private branch exchange or switch. They may have the exclusive use of private circuits to another terminal or private branch exchange, bypassing the common user network with a "dedicated" or "private" network. Individual circuits in a dedicated network are likely to be more heavily loaded than those provided between individual terminals, but will still be less economically utilized than those provided in a common user network. The relative advantages and disadvatages of meeting needs by common user and dedicated networks will be discussed in a later chapter.

9

MAIN TELECOMMUNICATIONS NETWORKS

In a common user system the circuits between switches or exchanges are usually regarded as the main network, in contrast to the circuits between the terminals and the switches, i.e., the local network. Circuits in the main network are longer, sometimes transcontinental or even transworld. In designing and providing the main network the dominant considerations are transmission standards, control signalling, and cost. The longer the circuit, the greater the attenuation and distortion of the communication being carried, and the greater the attention which has to be given to maintaining its quality and strength as it is being carried to the distant terminal. And the longer the circuit, the more attention must be given to the ways in which its use is controlled.

In the local network the distance between terminals is often short enough for the communication sent by one terminal to reach the other with sufficient clarity and volume so that no amplification or other correction is needed en route. Often direct current signals can be used to control the use of the circuit. For example the required number can be signaled by sending out direct current pulses from the terminal telephone, and the end of a telephone call can be signaled by switching off a direct current. In the main network, without amplification and other correction the communication being carried would be unintelligible when it reached the distant terminal. The resistance of the circuits is too high to the passage of direct currents for these to be used. The costs of the cables or radio systems to bridge the distances covered by main networks are not only high, but are further increased by the transmission and signaling technology which has to be employed.

The very high costs of a main network can only be adequately supported by achieving a very high level of usage. (The provision of circuits dedicated to the use of single terminals or small groups of terminals is difficult to justify for other than very high levels of utilization of that one terminal or group.) For that reason telecommunication administrations attempt to raise the loading of main networks in various ways and to increase the communications carrying capacity of the circuits. They also, by varying charges, seek to even out the load on the network, eliminating peaks and troughs. Through the switching arrangements and the routing of traffic the load is concentrated and in that way utilization increased.

By dividing the band of frequencies which can be carried by a single circuit into a number of separate communication channels the carrying capacity of circuits is multiplied. In some instances the use of a channel is shared over time by several terminals, and the time the channel is actually carrying communications is thereby increased. Most providers of main networks offer incentives to users to shift their communications to those times of the day or week when the network would otherwise be underutilized, thereby reducing peak demand and taking up network capacity which would otherwise not be earning revenue.

MAIN NETWORKS CONSTRUCTION

The earliest long distance circuits between switching centers were provided by overhead wires carried on poles running along the sides of the main roads between towns, each pair of wires carrying one telephone or telegraph circuit. To keep resistance as low as possible, very heavy gauge copper wires were used and very strong methods of construction were needed. The poles had to be large, firmly set in the ground, and securely stayed. The construction and maintenance costs were high, and there was an obvious physical limit to the number of circuits which could be provided between towns using such methods.

The overhead routes were replaced by underground cables as soon as the number of circuits needed were sufficient to justify the high initial costs of digging up the roads, laying earthenware ducts to contain and protect the cables, and constructing jointing chambers to join together the lengths of cable. The high initial costs were further increased by the need to provide "repeater" stations at relatively frequent intervals for amplifying, or restoring the strength of the communications being carried and for correcting any distortion.

Inevitably, given such high construction costs, attention was directed to finding ways of increasing the number of channels which could be carried by a single pair of wires. In various ways the number was increased, initially by a method called "phantom working" ("phantom" because additional channels were made available without increasing the number of wires in the cables). From this a method of working called "carrier" was developed which ultimately made it possible for up to thirty channels to be carried by a single twisted pair of wires.

With the continued high rate of growth in the demand for telecommunications, other ways of increasing the number of channels between switching centers had to be found to avoid constantly having to provide additional cables. The new coaxial cable used an entirely different method of cable construction. Instead of having some thousands of separate pairs of wires, a coaxial cable has just a single stout copper conductor running down the center of a flexible copper tube. In the larger cables several such tubes are contained in a single sheath. Such cables can carry a very wide band of frequencies which can be divided between several hundred separate channels of communication. The additional capacity is gained, however, at the cost of having to provide more equipment along the route of the cable to maintain the standard of transmission, and having substantially more equipment at the ends of the cable for channel frequency allocation and separation.

The next stage of development was the provision of line of sight radio routes. These, using extremely high frequency radio beams, are also capable of carrying many hundreds of separate channels of communication. The heavy costs of inter-switching center cabling are avoided, but radio stations have to be provided at frequent intervals and virtually within sight of each other. Environmental problems can arise, as the ideal locations are on the tops of mountains and other high ground, and a radio tower is regarded by many as objectionably unsightly. Radio networks can, however, be provided quickly and are particularly useful for short-term needs, for example a major sporting event that does not justify the laying of cable to provide communications for only a few days. Many broadcast radio and television services are dependent on the line of sight radio networks to relay programs to local transmitting stations.

In the last two decades we have witnessed the development of communications satellites. Essentially very high-frequency radio signals providing a broad band of frequencies are transmitted from Earth station to Earth station via satellites in stationary orbits above the earth. Initially satellite communications were used for bridging long distances and primarily as an alternative to submarine cables for intercontinental service. The capital costs were high and only by achieving a high load factor, and carrying communications for which users were prepared to pay relatively high charges, could the development be made commercially viable. To ensure a high load, intercontinental satellite systems usually only carry part of the demand with submarine cables being used in parallel. Costs are coming down and shorter-distance services are now also beginning to be carried by satellite, particularly to meet unforeseen and short-run demands, or where other means are impracticable or uneconomic; e.g., for communications with offshore oil platforms.

Satellite communications are primarily employed for common user services with charges based on the actual time a channel is in use. Radio and television broadcast authorities make extensive use of them for news and other topical programs. Some major organizations rent satellite channels for their exclusive use, but for the foreseeable future the number doing so is likely to remain limited.

One of the results of the convergence of telecommunications and computer technologies has been the use of digital techniques. It is possible to convert a current with a varying frequency and amplitude—for example speech—to a series of on/off pulses. These digital pulses can be converted back at a distant terminal to produce a current varying in frequency and amplitude in the same way as the original. Digital communications take up less of the frequency bandwith available in a cable or radio system, and are less susceptible to distortion and other transmission weaknesses. Digital techniques for voice and nonvoice communications can be carried in virtually the same way through the network and the utilization of the system is accordingly far more flexible. In the more developed networks the transmission and signaling paths may be separated to increase the speed of operation and working flexibility. Until comparatively recently switch design and main network design were approached separately, with the two interfacing in essence. Increasingly the two are being integrated more closely, with more attention being given to raising main network utilization through switching development.

Digital transmission has been used increasingly in cable, radio, and satellite networks, but has even greater potential in optical fiber networks. An optical fiber cable is made from very fine threads of glass, and communications are carried by a beam of light produced by a laser. A single thread of glass fiber has a capacity for many hundreds of simultaneous channels of communication. The cables are small in diameter, but in spite of this efficiency higher utilization is gained at the cost of relatively heavy terminal and other equipment costs.

VALUE ADDED NETWORK SERVICES AND LEASING OPERATIONS

Telecommunications corporations are able to earn a high rate of return on the capital invested in main networks. They do so partly by tailoring the size of their networks to demand, minimizing spare capacity, and achieving a high load factor. More important, users place a higher value on long distance communications and are presumably prepared to pay higher charges. Telecommunications corporations argue that such higher rates of return are needed to subsidize local services, whose charges often do not cover their operational costs. These corporations over the years have steadfastly refused to lease channels in their networks for other operators to rent in their turn to the public, possibly with enhanced facilities. They have maintained that they will meet any demand for facilities and have resisted any challenge to their monopoly of long distance communications. In some countries recent legislation has compelled the telecommunications corporations to allow operators to lease channels in their networks for rental to the public. In Britain the leasing arrangements are on condition that the service provided thereby offer facilities not otherwise available; i.e., value added services must be provided.

In the United Kingdom the relaxation of the past monopoly of the corporations has yet to result in the introduction of many new value added services. The corporations themselves have possibly been provoked by the competition to be more enterprizing in the range of standard services they offer. In the event possibly the demand for value added services is not so great as had been assumed. It is possibly in the nonvoice field that value added services have the greatest potential, for example in providing inter-connection between terminals which would otherwise be operationally in-compatible or in offering higher operating speeds. Certainly the erosion of the high rate of return on main network services which corporations claimed would jeopardize the cross-subsidization of local services has not yet become manifest.

MAIN NETWORKS SUMMARY

No more than a brief, highly simplified account is given above of the development and characteristics of main networks. In practice, of course, existing main networks are made up of a variety of methods of construction. They include open wire, overhead routes at one end of the spectrum to satellites at the other. With the growth in communications and the development of techniques to increase the number of channels which can be carried by a single pair of wires, coaxial cable, radio system, satellite, or optical fiber, the costs of carrying communications in a common user network have been brought down dramatically.

Initially and for many years the costs were closely related to the distances over which the communications were carried. The tendency today is for the costs to be more related to the number of switching stages involved in setting up a connection between two terminals than the distance between them. Telecommunications corporations, however, are very dependent on the charges which they can demand for long distance communications to offset the costs of the less profitable or even loss-producing local communications. Fortu-nately users tend to value long distance communications highly and the market will bear the higher charges levied on them. This is a critical issue in the regulation of network services and the implications will be discussed in a later chapter.

III

TELECOMMUNICATIONS STANDARDS OF SERVICE

10

STANDARDS OF SERVICE

Reference has already been made to the quality of the communications transmission when it reaches the distant terminal, but that is only one of the many criteria on which users will judge the quality of the service given by a telecommunications system and on which user satisfaction depends, whether a common user or a private dedicated system. A substantial part of the capital and operating costs of a telecommunications system are incurred in ensuring that the received communication is loud enough, is relatively undistorted, and is not masked by background or extraneous noise. A telecommunications company may charge more than its regular rate to provide communications higher in quality than its normal standard. The additional costs of meeting a specific level of performance can be high. Usually such arrangements are based on detailed specifications, firstly to ensure that the user receives what he is paying for, and secondly so that the vendor can be sure that the user is not getting more in terms of utilization than he has paid for. (Data communication managers understandably have sought to use communication channels at higher speeds than the vendor intended or on which it has based its charges.)

The costs of providing and maintaining a communications system, whether a public common user system or a private dedicated system, depend to a large extent on the standard of performance required from it and produced by it. Costs and the standards of service are closely related. Failure to oversee and control service quality on the one hand can inflate costs, and on the other can lead to the service not being up to the standards the user is paying for and is entitled to expect. Standards of service are discussed in following chapters under the criteria of speed and quality of provision and connecting, availability, reliability, repairs, "security," support services and user relations, billing accuracy, and future hardware developments.

It is self-evident that if the user is assuming full responsibility for procuring and operating his own system, he should be responsible for ensuring that where appropriate the above criteria are controlled. With a common user system the vendor is responsible for service standards and for satisfying users renting its services that it is providing what they are paying for. Seldom in practice are many of the standards of service which a vendor is charging for specified. Seldom in practice are the standards of service to which a private dedicated system is to be operated either stated before it is procurred, or monitored or controlled subsequently.

There are three separate but interrelated stages in controlling the performance of a communications system: stipulating the standards to be achieved, monitoring and measuring actual performance, and taking corrective action to achieve and maintain the stipulated standards.

The relevance of these stages will be discussed under each of the above criteria. Essentially in stipulating standards the relationship between service and costs must be kept rigorously in mind. *There are no absolute standards of service.* The optimum standard of service is that which results in the level of operational difficulty or inconvenience which the user can just tolerate. If a higher standard than that is provided then costs are being incurred which are operationally unjustified. There may be other reasons still for incurring them which cannot be objectively and financially supported, but they should be recognized as such. For example, corporations may be concerned not to attract undue attention from governments or regulatory authorities: Corporations with a monopoly or quasi-monopoly of common user telecommunications may consider it expedient to spend more than would otherwise be justified to avoid any significant level of complaint, recognizing that political reaction to complaint about the services it is providing might outweigh any objective cause for concern. Similarly, the management of a private telecommunications system facing the task of obtaining the approval of its board for major investment in new equipment may deem it prudent to take exceptional measures to ensure that the services it is providing are not the cause of complaint, just at the time attention is being focused upon them.

If, on the other hand, a lower standard than the optimum is provided, then any immediate economy is likely in the longer term to be more than offset by penalties. For example, it might be thought that too many operators were being employed to answer incoming calls to a firm, and that incoming calls could be left unanswered for a time. In some instances that could lead to business going elsewhere because callers were not prepared to wait for attention.

Penalties following attempts to reduce expenditures by lowering standards of service may not become immediately apparent, but can jeopardize performance long after the economy has been made and when possibly the short-term pressure for it has passed and been forgotten. It is often the case that standards of service can fall or be brought down far more quickly than they can be raised, and that their restoration can require far more expenditure than was saved in the first place. Standards of service have to be striven for over an extended period and once achieved should not be sacrificed lightly.

11

SPEED OF INSTALLMENT

DISTRIBUTION OF LINES

When an installation is being provided for the first time to a new house or office it is usually known well in advance when it will be required for use. Provided that notice is given of the starting date as soon as it becomes known, the telecommunications company or suppliers and installers of the terminal equipment should be able to plan and allocate resources to ensure that the installation is available for use as soon as it is wanted. Close cooperation between the developer of a new building site and the telecommunications engineers remains essential, however, to ensure that any underground cabling is installed before roads and footways are completed. If underground cabling right into the new buildings with no overhead distribution is required, provision must be made for this when the foundations are being laid. Similarly within the building all cabling should be completed before flooring and walls are finished, and wherever appropriate, provision should be made for inter-floor cabling. Such obvious provisions are made for drainage and other services, but often not for the distribution of telecommunications services.

Failure to have an installation ready on time can be the fault of the user in not giving sufficient notice or in changing requirements often and up to a late stage. Too often telecommunications requirements are only decided when the construction of a building is well advanced and inadequate provision has been made for internal wiring and even the entry of cables into the building.

Telecommunications requirements must be determined in advance; detailed specifications of those requirements must be given to the telecommunications company, suppliers, and installers at the earliest possible stage; and a member of the management commissioning the new building must be made clearly responsible for seeing that those requirements are met. This will be returned to in a later chapter.

The telecommunications company or other provider of the local external network in the past has been faced with a difficult dilemma. The local external network may represent between 25 and 30% of the capital invested in the system. On average, however, this will only be earning revenue about five minutes in twenty-four hours. There is, therefore, understandable concern not to provide more cable pairs than will be used. On the other hand, if the margin of spare equipment is kept too low it may not be possible to provide service when it is required. There are ways of marginally increasing the utilization of the local network by sharing the use of a pair of wires between two users or adopting other "pair gain" techniques. These are at best temporary solutions, not final solutions, for meeting the requirements of businesses. The problems the provider of a local network faces are further compounded in that although a need may be foreseen in a particular district, the precise location of future installations cannot always be predicted. A solution is sought in making the final distribution as flexible as possible. The underground cables may be taken to the foot of telephone poles and distribution completed by overhead wires, which can be erected and taken to wherever service is required, relatively quickly.

A number of the branch cables from the poles are terminated at a "flexibility point" where pairs of wires can be cross-connected to pairs in the main cable running back to the switching center. Not all the pairs in the branch cables normally will be in use, and only those that are, are connected to the main cable pairs. The total number of pairs in the branch cables exceed those in the main cable, and are cross-connected as and where required. In this way pairs running back to the switching center can be made available for cross-connection to a far wider area than would otherwise be possible.

The speed of provision of initial service is dependent in the first instance on the availability of cable pairs to connect terminals back to the switching center. At times of a high growth rate companies have not always had adequate pairs available, or available in the right places, due in part to a shortage of capital and to cable manufacturing delays. The dilemma remains, however, for local network providers of achieving a high level of equipment utilization while at the same time being able to respond speedily to unforeseen growth in a particular locality. Users have often failed to understand that unlike electricity distribution where, within limits, additional users can be connected to a common supply, distribution in a telecommunications system requires that each terminal have its own dedicated pair of wires back to the switching center.

The need to have pairs of wires available in internal local networks to connect terminals to a private switching center is equally important. Although shortages can probably be corrected more quickly and the diseconomy of excess provision of line equipment is less serious, if a good speed of provision is to be achieved and a building is not to be festooned with unsightly wiring, the careful planning of internal local networks is equally important.

INSTALLATION OF TERMINALS

The actual time required to install a telephone or other terminal equipment is relatively short. Providing the line equipment—external or internal—is available, generally it should be possible to have an installation ready for use as soon as the user requires it. In practice this is not always the case and installation delays are a major cause of complaint. Businesses which may be forced to move early to new premises or to make other accommodation changes are particularly critical of installation delays. Dependant on telecommunications, their operations can be seriously jeopardized by failure to have installations ready on time. Failure to specify precisely what is required and to have accommodations ready for the installation can be the reason for this, and again underline the importance of firms making a manager clearly responsible for their communications. If the requirements to be met are not specified or are changed when the installation engineer arrives, the telecommunications company cannot be blamed for any delay. It is particularly important with terminals which incorporate software programmed for the individual user's needs that requirements are stated clearly in advance of installation. It hardly needs saying that if the premises are not ready for occupation it may not be possible for the installation engineer to go ahead with the work.

More often the reasons for delayed installation are, regrettably, the responsibility of the telecommunications company. The installation engineer may find the line equipment allocated to the particular installation not available or otherwise unusable. The particular terminal ordered may be out of stock and service may have to be given initially using a temporary expedient. In an attempt to use its installation manpower most productively, the telecommunications company may make inadequate allowance in its work schedules for unforeseen delays and fail to meet the installation dates it has given. Installation managers understandably try to avoid unproductive traveling time between installation orders and seek to arrange jobs in an order which makes for an efficient deployment of their labor forces but pays less heed to customer convenience.

Rightly or wrongly, users expect installation orders to be met promptly, albeit sometimes they do not need to use the installation as soon as it has been provided. Speed of installment is one of the criteria on which the service of a telecommunications company—or the communications management of a firm—is judged. Telecommunications managers in this respect face a dilemma, as they do in many other aspects of their work. They have to balance the importance which users place on speed of provision against the lower productivity and higher costs incurred in meeting users' wishes.

12

STANDARDS OF INSTALLATION

Users judge the standards of installation on the following criteria:

1. Whether the initial reception of an order for service was dealt with in an efficient manner—including being effectively advised on the products and services that best meet his needs.
2. Whether the installation was ready for use when required (even on relatively short notice or after a changed date.)
3. The length of the delay if the installation could not be provided on the required date.
4. The manner in which the work was carried out.
5. Whether the installation functioned as it should have after installment and whether the equipment and service was as specified.
6. Whether the charges were reasonable.

A number of these criteria can be assessed by taking objective measurements. The promptness with which installation staff respond by telephone or in writing to installation requests can have an immediate and critical influence on users' perception of installation standards and can be monitored. (Delays in obtaining an answer to telephone calls or a response to written orders create uncertainty and lack of confidence from the outset.) The extent to which it is not possible to immediately meet the customer's order because of lack of line equipment or terminal equipment can be measured. Delays are not only critical in creating customer dissatisfaction, but critical also from the telecommunications company's point of view because delay in meeting customers' requirements represents potential loss of revenue. Finally, the incidence of service failure on new installations after commissioning can be measured.

Customers' perceptions of installation standards are also formed, however, by a number of other and more subjective factors, the strength of which can only be measured by seeking their views in response to a questionnaire or other form of opinion polling. As with all opinion polling, the methodology must be carefully developed if the results are to be representative.

Objective or subjective measurement of standards is only a first step. Distinction has to be drawn between relatively straightforward installations requiring limited resources to provide, and the more complex requiring greater engineering skill and possibly more time to process and commission. It is also essential to distinguish between average performance and the service as perceived by the individual, albeit prejudiced, customer. Many thousands of orders may be dealt with daily by a telecommunications company, and of these the majority may be met at the time and in the manner required by the customer. But it is the comparatively minute percentage of orders which are not met on time or in the manner requested which will create a poor image for the company. All service measurements should, therefore, be expressed not only in simple arithmetical averages but also in terms which show the extent to which stipulated standards were not met and the distribution of the substandard occurrences.

As in the United Kingdom and elsewhere where the monopoly of the telecommunications administrations is eroded and increasing competition is faced, greater concern to create a good reputation for the prompt and efficient acceptance and fulfillment of orders for service may be expected. Equally important, as businesses come increasingly to manage their own communications more directly, they should increasingly monitor the performance of communications suppliers and assess their competitive merits. This will also identify any weaknesses in the firms' own communications management jeopardizing availability of communications.

The measurements taken may vary in detail, but basically should cover the following:

1. Percentage of orders for service met on the date requested.
2. Actual orders not met on the date requested with the number of days late in each case.
3. A list of orders still not met after the date requested, with the number of days outstanding in each case.

4. A list of orders on which there were service failures within a year's time after commissioning.

Such measurements provide telecommunications companies with evidence of the standards being achieved in the face of competition, and focus attention on weaknesses needing early corrective action. When such measurements are taken by the communications management of firms they will also provide evidence for performance evaluations of service providers and the basis for resultant proposed improvements or changes, as well as identify weaknesses under the control of internal management.

As in all service industries the installation standards achieved will depend largely on the resources committed and the costs incurred. Companies have to decide what value they place on their reputations *vis-a-vis* those of their competitors, or the extent to which they are prepared to face public criticism. The fact remains, however, that given the large volume of installations orders placed with public telecommunications corporations, high standards should be attainable without incurring significant additional costs or depending on installation work being carried out in less time. It does require, however, close control of the allocation of jobs and deployment of the installation labor force to prevent a backlog of orders building up in the system with all orders being delayed awaiting their turn.

13

AVAILABILITY OF COMMUNICATIONS

Shoppers in a supermarket accept that at peak periods they may have to wait in line until a checkout assistant is free. Again, motorists are prepared to wait for petrol until a pump is free. In both instances they can see the reasons for the delay, and they accept that it would not be reasonable to provide so may checkout points or petrol pumps that even during the busiest periods they would never have to wait for an assistant or a pump. They are less inclined to accept that their telephone calls or other telecommunications may be delayed for similar reasons. This is partly attributable to their inability to see that all the operators on the one hand, or all the switching equipment or lines on the other, are in use; leading them to assume that their needs are not being properly met. It is also partly attributable to the sense of immediacy which the use of telecommunications generates and which creates the lower tolerance users display when for any reason a connection is not immediately available.

Again a balance has to be struck between the costs of having enough operators or switching equipment and lines that connections are never delayed, against allowing connections occasionally to be delayed. Connections set up via an operator will be delayed, the caller listening to a ringing tone, until an operator becomes free. Connections via the switching equipment which cannot be immediately established with one exception actually fail at the first attempt and the caller has to try again later. The exception occurs right at the start of an attempt to set up a connection. Connections may be delayed at the start until a first switch in the train of connections becomes free, delays normally unnoticed by the user, who is unaware that the dialing tone was not received at once.

SWITCHBOARDS

If the number of calls to be handled and the average time taken by operators to establish the required connections are known, it is possible to calculate the number of operators required to ensure that a prescribed percentage of calls are answered within a stipulated time. For large switchboards with a high level of traffic from a large number of users who are initiating their calls at random, the calculation provides a sound basis for assessing the number of operators required. But it must still be recognized that the calculation is based on forecasts of the number of connections that will be required and when the peak period will be. The forecasts may be wrong and the percentage of calls waiting the stipulated time may be greater than prescribed, or conversely be lower. The operators might not be working at the required level of productivity, or the peak period could occur at a different time than expected.

External factors may change the incidence of users' demands for connections. Political, economic, and even climatic uncertainty can cause a sudden rise in the use of telecommunications not foreseen or allowed for in the staffing of switchboards. Sudden fog disrupting air, rail, and road travel can cause demands on switchboards to rise markedly above the forecast level. It is at such times, when users need to make urgent calls, that they are likely to be even less tolerant of delays. Even a change in the timing of a particularly popular television program—or sometimes a change of fortunes for England in a Test Match—can cause a distortion in the incidence of telecommunications usage.

The reliability of the forecasts of operator requirements depends critically on the average time an operator takes to set up a connection. Provided the type of connection being set up is simple and requires limited attention by the operator, the staffing will line up well with demands on the switchboard. If, however, there is a wide variation between connection times, there is a strong possibility that at times all operators will be engaged on connections that are taking a long time to establish and other callers will encounter delays. The average time taken by operators depends not only on their efficiency but also on callers themselves. They can cause delays by not passing their demands in an efficient way and taking up the time of the operator in obtaining from them precisely what their needs are. For this reason, it may be difficult to get through to the directory inquiry bureaus. Callers often have the vaguest information about the firms or persons whose numbers they require, not even knowing how the name is spelled or what the address is. The time taken by the operators in clarifying callers' needs can be prolonged and can well exceed the average time taken to handle inquiries. At times, all the staff at the bureau can be busy dealing with such inquiries, and callers are inescapably kept waiting.

Achieving the required standard of service from a switchboard depends to a large extent on a supervisor or senior operator coordinating the work of the team: observing when callers are being delayed and taking corrective action; if necessary bringing additional operators to the switchboard if that is possible. The standard of service given by a private switchboard depends to a large extent on the users recognizing that the primary role of the operators is to establish connections. If the operators are required to locate individuals, to look up numbers in telephone directories, and to act as message carriers, then it is probable that callers will encounter delays at times. If these are external callers such as customers of the firm, they are unlikely to be impressed.

The standard of service given by a switchboard can best be measured in terms of the percentage of calls unanswered in say five, ten, and fifteen seconds and any going unanswered for longer times being individually recorded. Again, the critical factor is not the number of calls answered within the set times but the number delayed longer. While the delay rises linearly, criticism rises exponentially.

SWITCHING EQUIPMENT

It would be quite uneconomical to provide sufficient switching equipment to ensure that a path for a connection through the system was available at all times. A balance has to be struck, as in the case of operators, between the costs of providing equipment against the consequences of some failed attempts because there is no free path available. There are a number of significant differences between striking the balance in the case of the operators and in the case of switching equipment.

The operator is only engaged when setting up a connection: Once it is established he or she is free to accept another call. Later the operator may have to return to the connection to record the charges and to disconnect it, but that can normally be done as an overlapping task, and on the latest switchboard positions even that degree of further involvement is unnecessary, as charges are recorded and connections are terminated automatically. In contrast most of the switching equipment needed to establish a connection remains in use until it is released by the caller. A further major difference between operator-assisted calls and automatic equipment is that in the case of a connection set up by the operator, if the demand cannot be immediately accepted the caller is able to wait until an operator is free to deal with him and he need not call again. If all the switching equipment needed for a connection is engaged the caller receives a busy signal and must make a further attempt to establish the connection.

While the standard of service in the case of the operator can be measured in terms of the time callers wait for their demands to be attended to, in the case of the switching equipment the standard is expressed in terms of the proportion of failed attempts to set up a connection. Given the number of connections able to be carried in a given period, the average duration of a connection, and the specified proportion that will fail due to all switching equipment being in use (e.g., 1 in 500) it is possible to calculate the quantity of switching equipment required. Again in practice the calculation may not achieve the required standard of service.

The number of connections and the average duration on which the calculation is based is a forecast and may in actuality prove too low. An attempt is made to get greater accuracy once a system is up and running by measuring the amount of traffic flowing through the switching equipment in terms of Erlangs. (An Erlang is equivalent to that amount of traffic carried by one circuit continuously engaged.) Measurements are then taken over a period of time and attempts made to detect the pattern of growth and to forecast future requirements. To be successful there must exist a degree of underloading: any overloading will give a false measurement in that it will exclude attempts to set up a connection that fail because of an inadequacy of equipment. If the system is heavily overloaded users will be inclined to make repeated attempts without leaving time for the transient peak-loads to pass. Their repeated attempts will further aggravate the overload on equipment and lines preceding the point in the system at which the congestion exists. Severe overload on the system, in forcing calls away from the peak period, may lead to the real peak load being underestimated. Relief measures may not immediately prove effective, merely reveal suppressed demands requiring further equipment or augmentation of lines. Once a severe overload develops it may take some time and several attempts before the proper balance between equipment and line allotment and demands for connections can be reestablished and the required standard of service restored.

A further major complication remains of deciding the amount of switching equipment needed to achieve a specified standard of service. Unlike operators, switching equipment for all practical purposes cannot be brought into use and then taken away again when not needed. The employment of operators can be varied according to the varying load on the switchboard, subject to acceptable rostering arrangements. Switching equipment once provided remains as a capital expense whether in use or not. Switching equipment is provided for the busiest periods albeit for much of the time it will be standing unused. This places another limitation on the amount of equipment provided. It is the normal practice to provide for the busier periods but not for the busiest, and then to specify what is in effect a double standard. For example it may be decided to provide equipment on the basis that only 1 attempt to set up a connection in 500 should fail with the forecast level of traffic subject to not more than 1 attempt in 100 failing if the level of traffic is 10% higher than forecast.

There are no absolute standards: It is for individual companies to decide what its users will accept as reasonable *vis-a-vis* the charges they will have to pay. Companies attempt to influence demand and to stimulate a more even traffic pattern. One obvious way is to have different rates for different times of the day or different days of the week. If this proves too successful and the peak load subsequently develops at a time when charges have been reduced, resulting in decreased profits, then a lower standard of service may be applied and the amount of switching equipment required accordingly reduced in order to save money. (Off-peak telephone calls tend to be less urgent and callers accept the marginally lower standard of service.) Attempts should be made to stimulate additional use of the system when the equipment would otherwise be lying idle.

It is doubtful whether the promotion of additional off-peak usage of telecommunications is any more successful than it is for other public utilities. Although users who need to use the system may be persuaded to delay doing so with the attraction of cheaper call rates, they appear to increase total usage only marginally in response to promotion campaigns. What success such compaigns have appears to be short-lived and does not appear to establish a sustained higher level of usage. As such they may make little contribution to taking up otherwise unused system capacity.

MAIN NETWORKS

Similar standards can be applied to determine the number of channels needed between switching centers. Forecasts are made of the number of simultaneous connections required between switching centers in the busier periods and their average duration. The number of attempts to set up a connection that fail due to all channels being engaged is specified and the required number of channels calculated. The same limitations apply as in the case of determining switching equipment, but higher failure rates are normally allowed on the longer and more expensive routes. In some instances arrangements are made for alternative paths through the network to be used automatically if the primary routing is fully loaded and free channels exist on the alternative routing. In this way not only does the user obtain a better standard of service, but the overall loading of the system is raised.

OVERALL STANDARD

The various stages of switching equipment at each center are individually dimensioned to achieve stipulated connection failure rates, and the various routes between switching centers are similarly dimensioned. The chance of an attempted connection failing due to all equipment being engaged is thus compounded. It is customary for telecommunications companies, therefore, to monitor not only the loading of every switching stage and route but also the overall failure rate being experienced by users. Again, however, it is not the average degree of failure due to system overload that is likely to concern users, but the individual number of occasions they fail to set up a connection on the particular paths through the system that they most frequently want to use. Their perception of the overall standard of service thus at times may be worse than that observed by the company. Almost inevitably the companies are compelled to operate with comparatively lightly loaded equipment in order to avoid criticism on this score.

The user will in practice, of course, encounter other causes of failure to set up a connection. A fault may exist in the system itself, preventing the required path from being established; this will be discussed later. The user may fail to operate the terminal correctly or may make a mistake in selecting the required number. (There is some evidence that selection errors are made more frequently with keying than with dialing.) The required distant terminal may either be busy or unattended and incoming calls may be going unanswered. Users tend at times to attribute failures to set up connections to system failures when the real reason lies in their own misdialing or in the required numbers not being available to accept their calls at the time they want them.

System failures should be no higher than 0.01% of total attempts to set up connections, and failures due to switching or line equipment being engaged should be no more than 1.0%. In contrast failures due to user misdialing, or due to the required number being busy or unattended can be of the order of 20% of total attempts. The precise figures will vary from system to system but the relationships will be of that order.

Telecommunications companies cannot disregard failures due to misdialing or the required number being busy or unattended. Such failures engage the equipment while the user makes the attempt, but are nonrevenue earning. Equipment has to be provided, in effect, to enable the user to make an unsuccessful attempt. Further, as mentioned above the company may be wrongly blamed for such failures. Attempts to minimize user mistakes by providing instructions with terminals, offering (sometimes free of charge) instruction courses for users' staff, and constant exhortation to use the system correctly all help but in practice a degree of misuse is inevitable. With electrical and other domestic appliances the user can be made aware of the consequences of a mistake. (The coffee percolator will soon indicate if it is being used without putting the water in first.) The user is not made immediately aware of misuse of telecommunications equipment.

If a telecommunications company finds that connections to a particular number are frequently failing because it is busy, it will attempt to persuade the firm concerned to rent additional lines. Not only does it want the additional rental but also to reduce the incidence of call failures for which it is earning no revenue. A common cause of a high incidence of attempts to a particular number failing due to its being busy is an inefficient use of the installation. Lines may be engaged by other incoming calls going unanswered; incoming calls may have been answered but the caller left waiting until an extension is answered or someone returns to the telephone. Once an incoming call is answered the telecommunications company can begin to collect revenue and to that extent its interests are being met, but clients and customers of the firm concerned are unlikely to gain a good impression of it from such poor telecommunications operating practice. Some telecommunications companies have the statutory power to require additional lines to be rented if a firm is causing excessive failures due to all its incoming lines being frequently busy.

It is not in the interest of a firm to have all its incoming lines constantly engaged, as potential customers might well become frustrated and take their business to a competitor. Even when the management of a firms's outgoing communications is reasonably good it is often the case that little attention is given to the quality of its service from the point of view of users attempting to call in.

14

SPEED OF
CONNECTION

The speed with which connections can be set up depends not only on the availability of operators, switching equipment, and line equipment and cables, but also on the time taken for the system to respond to the users' demands. On connections set up via an operator, the time taken by the operator to accept the demand for a connection; the accuracy with which the demand is recorded; and the operator's competence in setting it up determine the speed of connection. The effectiveness with which the user responds to and works with the operator is also a major determinant of speed of connection. The effect of the user passing his demand in a concise, efficient manner and an equally efficient operator response is exemplified by the communications between airline pilots and air traffic controllers.

With an electromagnetic dial-controlled system, the time taken by the user to dial the digits of the required number is usually longer than the relatively slow switches need to respond. When the dial is replaced by a key pad and the user keys the required number, he is able to do so much quicker than the electromagnetic switches can respond. The delay between completion of keying and the setting up of a connection becomes very apparent and irritating to users. The advantages of keying are only fully realized with electronic fast switching systems, which have a faster response time than even the user with a key pad can match. The longer setting-up time with an electromagnetic system is even more marked if the digits of the required number are stored and translated into others for route selection. In extreme cases the delay may be so long that users abandon their attempts prematurely and before the time needed to set up connections has passed. The telecommunications companies may regard such failures as attributable to user error, a view that users find difficult to accept. With keying and electronic systems the speed of switching is such that not only is it faster than the user can key, but there is sufficient time for an alternative path through the system to be selected if for any reason the primary path is closed. Thus the selection of an alternative path is so quick that the user is unaware it has occurred.

A problem for telecommunications companies is that users' expectations are being raised with the introduction of fast switching electronic equipment, but this equipment is being installed piecemeal into the system. A firm may have, for example, an electronic private branch exchange and users who are accustomed to fast speed of connection internally. The speed of connection in the public network in contrast will be determined by the performance of the slowest link. If this is via an old electromagnetic switching center, users will experience a noticeable delay after keying before the connection is set up. Their frustration will be aggravated if the local public switching center is electronic and a fast speed of connection is experienced on local calls. The delay on long distance connections still passing through one or more electromagnetic centers will be that much more apparent and difficult to tolerate.

15

QUALITY OF CONNECTION

The standard of service also depends on the quality of the connection after it has been established. For voice communications there are three basic criteria: the strength of the received speech, the extent to which it is distorted and does not faithfully reproduce the sound as it was transmitted, and the extent to which noise and other extraneous sounds can be heard. For nonvoice communications the speed with which signals can be carried is also an important criterion.

Telecommunications companies again have to balance the cost of achieving high standards of performance against what users will accept and pay for. For voice communications, lower standards are set in terms of the frequency band width provided for and are generally regarded by users as acceptable. Users are inclined not to expect the quality of the received speech from a distance to be "high fidelity." Their expectations are rising, however, with experience of the better quality of speech on intercontinental connections via satellite or on the latest submarine cables. They are increasingly inclined to compare the lower quality of speech on local connections with that received from the other side of the world unfavorably.

Ensuring that the strength of the received signal is adequate is normally well taken care of in the engineering of systems, and in the care taken to restore the strength of signals as losses occur. A more critical requirement is to exclude noise and other extraneous signals. Electromagnetic switches with many moving parts and contacts are a source of much of the errors that occur in transmitting data and the noise users experience on telephone calls. Again, the contrast with fully electronic systems is most marked and focuses attention on the relatively poor performance of the older systems.

The detection of causes of noise and their elimination is one of the more difficult problems for telecommunications systems managers. It is often the case that the noise only becomes apparent when an end-to-end connection is established and in use. The source of the noise may be another connection and it may occur only intermittently. Elimination and prevention of noise depends to a large extent on the application of new technology and the replacement of the older obsolescent equipment prone to be noisy in operation. The quality of connections will improve but for some time yet users must expect the problem to be a continuing albeit diminishing one.

In the meantime demands for higher quality connections than the norm can be catered to with special engineering, but a premium will be levied by companies for the service.

16

RELIABILITY

As users have come increasingly to rely on telecommunications, they have understandably become less tolerant of breakdowns either in their own terminals or in the system itself. No matter how often they use the terminal or the system without trouble, if it is out of order or if it is not possible to set up a connection when they want to, it is that failure that concerns them. No matter how good the service previously, at that moment it is the failure that weighs heaviest in their perception of the service.

Telecommunications companies recognize that their reputation depends more on the number of times terminals and their systems fail than on the overwelming number of trouble-free connections set up by their users daily. They monitor failure rates intensively and constantly seek to improve the standard of performance. Faults on terminals are recorded in great detail and analyzed constantly to identify causes for which corrective measures should be found. Companies look particularly for any incipient and general deterioration and try to take corrective action before the standard of service goes into a marked decline. It is the practice of most companies to generate traffic artificially and to look for failures, possibly applying more stringent conditions than users do. They seek to identify failures and isolate the faulty equipment before its failure becomes apparent to users. Terminals and the local distribution network are periodically automatically tested from a remote control center, again seeking incipient causes of failure before users encounter difficulty.

Terminal failures are usually measured in terms of service-affecting faults per line per annum, broken down into main categories of plant, e.g., telephone, underground distribution cables, or overhead distribution cables. An average terminal failure rate of well below one service-affecting fault per annum is achieved. Again, however, the average is less important than the incidence of faults on individual installations. A particularly close watch is maintained for installations on which a number of faults are reported within a relatively short period, particularly if the same fault is reported by the user more than once.

Very detailed records are kept of all failures of the switching equipment and in the main network, and are analyzed to identify weaknesses that require corrective measures. All the equipments' functions are regularly tested, applying more stringent conditions than would normally be encountered operationally, to detect incipient failures before they affect users' service.

17

SPEED OF REPAIR

Users accept that it is not economical to provide a telecommunications system that never fails, impatient though they may be if a failure occurs just at the time they wish to make what they regard as urgent use of it. They are not so tolerant, however, if the speed of repair is slow and faults are allowed to go unremedied for any length of time. The performance of a telecommunications system is judged as much on the speed of repair as on the incidence of faults. Indeed, the image of a telecommunications company is often enhanced by the speed with which it restores communications after a major breakdown.

The speed of repair of faults on terminals and in the local distribution network depends on rigorous recording of all failures reported by users, careful diagnosis of the causes of the failures, and effective deployment and control of the repair staff.

The initial processing of the reports from users is critical. Users cannot be expected to identify precisely the nature of failures: Indeed, unwittingly they can mislead and delay repairs. The reception staff need to skillfully interrogate users and obtain as much information about the nature of the difficulty being encountered as they can. This requires tact to avoid further irritating a user whose opinion of the service may already have been jeopardized by the inconvenience.

The next stage is equally critical in determining the ultimate speed of repair. With the information given by the user the repair center staff attempt to identify the causes of failures and to locate where in the system they probably exist. Using testing equipment, they attempt to diagnose the reason for the failure, whether it is attributable to a fault in the terminal; is in the distribution network; is in the switching center, affecting the equipment associated with that particular user; or is in fact not a fault of the equipment used by that particular user, but is in the common switching equipment or network. Incorrect diagnosis of the cause of failure not only delays clearance of that fault, but reduces the speed of repair overall by wasting the time of the repair staff. For example, a fault in the underground distribution network wrongly diagnosed as being on the terminal telephone will result in a telephone repair man making an abortive visit to the premises, and only later will the need for underground repair staff be recognized. In carrying out the repair the emphasis should be on repair and not just on clearance of the cause of the failure. There is a temptation to get service restored quickly, not taking time to ensure that the fault does not recur. It is equally important that careful records are maintained of the causes of all failures, so that if a further fault does occur not only are the repair center reception staff aware of the situation and able to deal with the user accordingly, but diagnosis and control of repair the second time is tighter.

The relaxation in some countries of telecommunications corporations' sole right to provide and maintain internal networks and terminals can cause difficulties. If users have the right to acquire terminals (albeit subject to prior technical and operational approval) from suppliers other than the telecommunications corporation, responsibility for repair and performance can be divided. Unlike domestic electrical appliances, a fault on a terminal may not be immediately and clearly recognized as such. Connection failures can be caused by faults anywhere in the system, but the previous end-to-end responsibility of the corporations may no longer exist, and they may understandably be reluctant to accept responsibility for failures until certain that the cause is not on the terminals provided and maintained by an independent supplier. Similarly they are reluctant to accept immediate responsibility because failures might be caused by faults on internal networks provided and maintained by independent installers. Telecommunications managers obtaining terminals and internal networks from independent sources should take particular care to ensure that clear and firm arrangements are made for speedy diagnosis and responsibility for the repair of faults and connection failures, and that the service of their companies is not jeopardized by divided responsibility.

Telecommunications companies constantly measure the speed of repair, and communications managers should keep an equally close watch on the "out of service" time on installations for which they are responsible. Again, average performance figures are inadequate, and any fault not cleared within two hours of being reported should be identified and separately listed. It would not be unreasonable to set a standard of well over 90% of all service-affecting faults being repaired within two hours of being reported, with virtually 100% being repaired within twenty-four hours. Records should also be kept of all faults which occur again within say three months and of all installations which have three or more service-interrupting failures in twelve months. In each case the installation should be specially overhauled.

18

SECURITY OF COMMUNICATIONS

Users have always been concerned about the confidentiality of the telephone, even though they rarely show equivalent concern about their face-to-face conversations in public places being overheard. The use of the telephone is regarded as somehow more intimate than even face-to-face conversation. Indeed things may be said over the telephone which it is doubtful would be said directly. Any possibility that telephone conversations are being listened to by a third party is regarded as an infringement on the liberty of the individual.

Particularly with the growth of nonvoice communications, the security of communications is regarded by users to be of great commercial importance. Within a public switched network confidentiality of communications is high because a single connection is anonymous among many thousands of connections constantly being set up and terminated. Once the connection leaves the common user network, however, it can be identified readily and its security is accordingly placed in jeopardy. As telecommunications are increasingly relied on for carrying highly sensitive commercial data and other information, the security of communications has become a major criterion on which users judge the standard of service.

The right of the State to monitor telecommunications in the interests of national security or in pursuit of crime prevention or detection is a political issue which it would not be appropriate to comment on in a book of this nature. Possibly of equal and growing concern, however, are the illegal attempts by commercial competitors and others with criminal intentions to monitor or have access to telephone calls or nonvoice communications to obtain commercially sensitive or confidential information. Essentially there are two safeguards users can take.

Firstly, they can scramble or code the transmission of speech or data in such a way that it is unintelligible to anyone intercepting it. (As an ultimate precaution, sensitive or confidential information which is not time critical should be physically transferred so that it can be guarded while in transit. For example financial information that is not required immediately is better conveyed by private courier than transmitted over the telecommunications network if it is highly sensitive or confidential and time allows.)

Secondly, access to central data banks over a telecommunications network should be barred to anyone not provided with the appropriate access code. It must be remembered that access codes are not always kept secure by the users, and in the last analysis the only safeguard is not to provide access to a data bank from the telecommunications network.

19

ADEQUACY OF
SUPPORT
SERVICES

Users do not judge the standard of service only on operational performance:
The adequacy and standard of support services is also critical to the image a
telecommunications company creates for itself. As mentioned earlier, when
applying for service users need to be given information and guidance about
which installations and facilities available would best meet their needs. The
reception they receive when reporting service failures is critical. They also
expect support in other ways, for example when querying their accounts, or
when having difficulty using the system.

NUMBER INFORMATION

Telephone numbers, unlike postal addresses, are artificial and seldom
convey any geographical connotation or other information. Possibly someday
users will be able to indicate the terminal to which they wish to set up a
connection by keying its location or its owner's name—or even by merely
stating this vocally. For the moment, however, it is necessary to know the
number of the terminal to which connection is required. The standard of
service given by a telecommunications company is, therefore, judged by the

arrangements it makes to provide users with number information. Users decline to accept sole responsibility for making their phone numbers known, as they routinely do with their postal addresses. They expect to be able to obtain the numbers of the terminals to which connection is required from a telephone directory, and that in turn those wishing to communicate with them can obtain information about their numbers. The management of number information is a major concern and a significant element in the operating costs of a telecommunications system.

Number information is provided in three ways: by publishing directories, by operating a directory inquiry service, and by making arrangements to inform users of any changes made in the allocation of numbers. If these arrangements are inefficient the standard of service will be jeopardized. Inability to obtain number information inhibits use of the system. Incorrect information results in abortive calls, irritating users attempting to set up connections and incurring unnecessary call charges, and those being called in error. Between 20% and 25% of the entries in the published directory change between annual issues. Although calls to previously allocated numbers can be intercepted in various ways and then redirected to the newly assigned numbers, failure to provide changed number information can be a major cause of user irritation and complaint.

Unfortunately for telecommunications companies, the publication of directories and the operation of a directory inquiry service is not as straightforward as one would suppose. Users tend to be very concerned about the way their identities are shown in telephone directories or the way they are referred to by the inquiry operators, for reasons which have little to do with the operation of the telecommunications system. The private person is often concerned that his decorations and academic qualifications be printed in full, or that an address he regards as prestigious be given prominence. The business user often seeks to exploit the directory to advertise or otherwise promote his firm. Even the order in which entries are printed can be critical to a business. Directory inquiry operators have to be careful that they do not inadvertently favor one firm against a competitor. For example they are often asked for the number of taxi firms and must ensure, while being as helpful as possible to callers, that calls are not always directed to the first taxi firm listed. In contrast, some users do not wish their numbers to be known and the operators have to be equally careful not to reveal them.

Number information services are regarded by some telecommunication companies as an embarrassing liability from which they would like to be absolved. They seek to recover some of the substantial costs involved by publishing classified directories, listing firms by trade or professional classifications, and charging for additional and prominent listings and for advertising. Such directories provide a valuable source of commercial information and yield worthwhile revenue to the administrations, but they are more often used to find the addresses of potential suppliers of goods and services rather than to obtain number information. To a limited extent, the wishes of private users to accentuate their directory listings is responded to by offering specially printed entries for a supplementary charge. Some companies providing directory service charge for omitting entries from telephone directories and for arranging for their directory inquiry operators to withhold number information. They argue that the omission of an entry from the directory results in additional calls to the directory inquiry service where special precautions against disclosure of the number have to be taken, and that this should be charged for.

Some telecommunication companies also charge for directory inquiries. They argue that the costs of providing number information should be met by users in two ways. The costs of directory publication, after allowing for advertising and special entry revenue, are regarded as being covered by the rental for installations. They consider the costs of the directory inquiry service, in contrast, should be paid for by users in relation to the use they make of the service. As many as 50% of the numbers obtained from directory inquiries could be found in the directory already supplied to the user, but because the directory inquiry service is more convenient it is used instead. Companies providing directory services tend to regard this as an abuse of their facilities which they are justified in charging for, in the interest of users who properly restrict their demands on the service. Extension users often fail to use directories, asking their private exchange operators to look up numbers for them, and busy PBX operators in turn go to directory inquiries. In seeking to achieve a proper balance between the use of the published directory and directory inquiries, the problem is to ensure that those users who have a proper need to go to directory inquiries are not penalized, bearing in mind the 20%–25% rate of change of published number information. Probably no aspect of telecommunications excites more public interest and concern than the availability of number information support services.

Attempts have been made to reduce the costs of directory inquiries and to improve the service to users by applying information retrieval theory and computer techniques. The success of such systems depends on the amount of information about the required number that has to be input before it can be identified. For example if an inquirer is only able to give a common surname and an imprecise address, such as "Smith" and "London," the retrieval of the required number becomes impracticable. Even the number of the street could narrow the search considerably; add the initials and it is possible the required number will be found. Systems utilizing information retrieval theory and computer techniques have been developed for directory inquiry operators, who progressively input as much information as they can obtain from inquirers until a single number is identified, with a strong probability that it is the required one.

One company has provided users with direct access to such a system. They input information about the required number by means of a simple alphanumeric key pad, and if the system is able to trace it, it is displayed visually on the terminal telephone. The weakness of all such systems is that the small percentage of unsuccessful attempts to trace a single unique number is still too high for them to gain the confidence of users. If and when such systems are successful, it is interesting to speculate whether users will need to know the numbering of terminals within the system, or whether they will merely have to input the identity of the person or firm with whom a connection is required (possibly orally).

For the time being the published directory remains the prime source of number information. The number of directory inquiries falls sharply in the days immediately following the issue of a new edition of a local telephone directory, and gradually rises again as the information it contains becomes outdated and as directories are misplaced. This is recognized as the "sawtooth" pattern of directory inquiries usage.

ASSISTANCE

Other than obtaining number information, from time to time users need assistance in setting up connections or to inquire about some aspect of the service. For this purpose most telecommunications companies provide support assistance or general inquiry operators. Some, however, take the view that the costs of such services are not justified. They argue that there is adequate published information about the services and facilities available, and that users should be able to rely on that. If for any reason the user is encountering difficulty in setting up a connection, the normal system surveillance will already have detected the failure and remedial action will already have been taken.

Other companies regard such user support services as essential to ensure that the system is properly used and to maintain user goodwill if serious operating difficulties, e.g., a major system failure, should occur. The opportunity is taken, however, to also have the assistance/inquiry operators provide special services for which charges are made. For example some companies use the operators to provide message agency facilities, others to provide a nationwide freephone service (U.K.) or toll-free service (U.S.A.). Firms in the United Kingdom quote in their promotional material a freephone identity, e.g., "freephone window," and callers wishing to contact them call the local operator, ask for the firm by its freephone identity, and the operator connects them free of charge to the firm concerned. The United States has an identical system of "toll-free" numbers whereby the sponsoring firm pays for the service and the toll charges for all calls made to its number, even those which are long-distance.

PUBLIC PAY PHONES

Another support service which excites user concern and reaction is the public pay phone service. In the earliest days of telephony the public pay phone was used by those who could not afford a telephone of their own for occasional—and usually urgent—use. With the growth in the availability of the telephone, the pay phone is less the "poor man's telephone," and more the telephone which is used while the user is away from home.

Pay phone service can be regarded literally as a support service, but the dilemma for telecommunication corporations is how to finance it. The installations tend to be technically complex, are vulnerable to vandalism and criminal attack, and are subject to high service and maintenance costs. It is very difficult to cover these from the charges collected. The question the providers face is whether to subsidize the costs of the public pay phone from their other profitable services or only to provide and retain those which are individually profitable. The service is politically very sensitive and will be discussed further in a later chapter.

20

USER RELATIONS

There may be significant differences between the standards of service as measured by the telecommunications company and as perceived by users. Users either regard the standard of service to be less than the objective measurements of performance indicate, or vice versa. Whatever the objective measurements of performance, it is advisable also to systematically monitor users' opinions of the standard of service. The analysis and presentation of both the objective measurements of performance and the monitoring of users' perceptions of it must be rigorous. Mention has already been made of how misleading average results are. Any conclusions should create a sound basis for decision making and not merely serve cosmetic purposes.

It was stated in an earlier chapter that there are no absolute standards of service. It is up to the company to decide the correct balance between the costs of achieving specified standards of service and the level of user dissatisfaction that they consider they can tolerate. Users' dissatisfaction will depend on their subjective perception of the standards of service. This perception of the service will be inherently subjective and will be influenced by users' expectations, which will be influenced by what users read in the press, hear on the radio, see on the television, and are otherwise encouraged to believe about the company providing service. The degree of user dissatisfaction will depend not only on the actual performance of telecommunications systems, but also on the success companies have in countering public criticism, in creating a good public image, and finally in securing and retaining users' goodwill. Thus the importance of user relations in deciding the standards of service to be achieved.

Reference was made above to the importance of support service standards. The cultivation of good user relations more and more demands the appropriation of specific funds for that purpose. User expectations and perceptions, given the nature of the service, tend to be inevitably biased toward being critical. It is the common experience of telecommunications managers to be criticized for mismanagement, but on further inquiry to find that the criticism is based on hearsay and not on the personal experience of the critic.

A good reputation has to be worked for and will not come merely from providing a good standard of service. A critical determinant of good user relations is the response of the company to the individual user when he encounters difficulty, real or otherwise. It matters little how good the service is normally, if on the occasion that it fails to come up to the user's expectations, he or she is treated in a manner that conveys lack of concern to remedy whatever the trouble is. Whatever one's opinion previously, it will be undermined. The better the service is normally and the more infrequent the causes of difficulty, the more users are likely to be critical when they do occur. Companies find it difficult, given the scale of their operations with many thousands of accounts being maintained and many millions of connections being set up without trouble daily, to inculcate in those staff in contact with users a constant concern for good user relations. Staff may find it difficult to accept that users, who on encountering what they see as a failure of the system may not be as tolerant as they might be, must still be given every consideration. Whether publicly or privately owned, the degree of regulation and oversight of telecommunications companies by the State will continue to be influenced by users' perceptions and expectations of them.

21

CHARGING ACCURACY

Possibly no aspect of service quality excites user comment and concern as much as the accuracy of charging or billing. With systems where the charges for connections set up directly by users are recorded in chargeable units on meters in the exchange or central office, users inevitably are inclined to question the accuracy of the accounts. This has led to pressure for the charges on all calls to be separately recorded and itemized on the accounts sent to users. Whether this leads to increased user confidence in the accounting system and to fewer bills being queried, is open to question.

Systems generally are designed on a "fail-safe charging" basis, i.e., if the charging system fails, charges will not be metered and the users' interests will be safeguarded. Assurances that this is the case, and periodical examination of the charging arrangements by outside authorities seldom satisfy users, who continue to press either for meters on their premises which they can observe or itemized bills which they can check against their own records or recollections.

Telecommunications companies constantly check the accuracy of their charging systems; they have an interest not only in ensuring that their users are not overcharged, but also that they themselves do not lose income by undercharging! There are, however, other factors that bear on users' confidence in the accuracy of charging. Telecommunications companies argue, with some justification and convincing evidence in support of their contentions, that users' own records and recollections are seldom accurate.

There are other more difficult and complex aspects of the problem. The person responsible for settling the account often does not know who is using the installation and the extent of the usage. Telecommunications companies asked to verify the use of an installation by connecting equipment in the exchange which records when it is used and where connections are going to can be placed in an impossible position. They may observe the installation being used for purposes they feel would be improper for them to reveal to the account holder. They are forced in such circumstances to state that there is no reason to doubt the accuracy of the charging system, leaving the account holder dissatisfied. They cannot be involved in matters that are clearly the responsibility of the account holder, and usually decline to give more than the bare facts recorded on the equipment, although they may know more. For example, they are not willing to become involved in possible matrimonial disputes. This is a problem not only in relation to the domestic installation, but also to the business installation where employees are using the firm's telephones for purposes which clearly would not be permitted by the employer if known.

The charge rates are often complex and make it difficult for account holders to relate the charges debited in metered units with actual usage. It is not unusual for charges to depend on many factors: the distance over which a connection is being set up, the duration of the connection in time, the time of day, the day of the week, and sometimes the time of the year. In part this is the result of telecommunications companies seeking to use the charge rates to relate charges more closely to costs and to achieve a more even loading of the system. The complex rate structure is also a result of user reluctance to accept a broader distribution of the rate structure, inescapably leading to some users bearing a disproportionate part of the overall costs.

22

ENHANCEMENT AND DEVELOPMENT

It is quite certain that the telecommunications field is only at the beginning of a long period of technological change during which the terminals and facilities available for users will develop and change quickly. The public's opinion of the standard of service provided by a telecommunications company will depend on the response of the company to the opportunities to enhance and develop telecommunications.

If companies attempt to adopt a restrictive attitude, placing obstacles in the way of users wishing to change and upgrade their communications, they will not be highly regarded. They must of course continue to be concerned about the technical and operational integrity of the systems they provide. For example, it was explained earlier that electrical signals have to be sent between installations and the exchange to control the setting up and disconnection of calls. Those signals are complex and have to be extremely precisely produced. If the signals sent out by a piece of equipment do not conform to the precise mechanical standards stipulated by the telephone company, call failures will tend to be high. The caller will wrongly blame the company and, more important, waste network time in ineffectual calls. Other people may be inconvenienced by receiving incoming calls from the substandard installation. Again, if an installation is to be effectively used over a public system it must transmit speech to the required standards of volume and clarity. Companies must ensure that the installation of a terminal by a user that has been supplied by another company does not in some way jeopardize the use of their own system.

If, however, restraints are placed on the attachment of terminals that are compatible, merely because they are manufactured by competing companies, then users are likely to be strongly critical. Users are likely to be similarly critical of any attempts by a company to limit the use of its system when there are no operational or technical reasons to do so. This aspect of the management of telecommunications will be discussed in some detail later under the heading of "regulation of telecommunications."

IV

TELECOMMUNICATIONS PLANNING

23

THE NATURE OF
TELECOMMUNICATIONS
PLANNING

There is hardly any aspect of industrial, commercial, or social life unaffected by the development of telecommunications. With the continued advance of the technology and the explosive growth in new products and the service industry, commerce and society will become ever more dependent on tele-communications. But the determination of telecommunications needs, the design of systems, the procurement of the hardware and software and its commissioning remain essentially tasks with long lead times.

Telecommunications companies have recognized this from the earliest days and planned the development and growth of their systems well ahead. In contrast, until recently major users have failed to recognize the importance of advance planning of communications. To some extent they have not been encouraged to do so: They have often had little freedom of choice, being restricted to installations approved and supplied by these companies. Few saw a need for a private dedicated system relying on the public switched networks provided by the telecommunications companies. Increasingly, users are being given freedom of choice, and with the growth in nonvoice as well as voice services there is a growing recognition of a need for private dedicated net-works, and expenditure on telecommunications has become a major element in companies' financial budgets. Private companies as well as telecommunication companies, therefore, are seeing the need for planning their future procurement of telecommunications well ahead. Equally important, they also need to base their ongoing utilization of communications on soundly based plans.

Planning for telecommunications divides broadly into three phases: strategic, development and procurement, and utilization. Strategic planning is concerned with preparing for the telecommunications opportunities that are expected to exist five years or more ahead. Development and procurement planning is concerned with deciding the level and timing of investment in new systems, in enhancing and developing existing systems normally up to five years ahead, and in procuring and commissioning them when needed. Utilization planning is concerned with ensuring the telecommunications systems and other resources acquired are used in the most efficient way to meet communications needs. It is normally concerned with essentially short-term decision making, that is, up to eighteen months ahead.

The whole planning process rolls forward annually with elements of strategic plans becoming development and procurement plans, as with the long lead times, much of the investment contracts have to be placed early, while a new strategic plan is developed. As systems are procured, or developed and enhanced, the emphasis changes to utilization. In general the confidence limits of the forecasts and predictions on which planning is based narrow as strategic planning moves forward into development and procurement planning, and the latter moves forward into utilization.

24

STRATEGIC PLANNING

Telecommunications planning until a decade or so ago was almost entirely concerned with assessing the rate of growth of basic telephone services, and with the timing and pace of automation. Those telecommunications operations owned by the State or which were closely regulated by the State had also to take into account any overriding constraints placed on their investment, or the acquisition of foreign manufactured systems. Private concerns rarely saw a need for telecommunications strategic planning, relying on the public corporations to have foreseen their needs and to have equipment available as these developed.

In recent decades the importance and complexity of strategic planning has grown immeasurably for both the public and the private concern. The range of services either existing or possible has widened, and the range of existing or latent technological options is constantly growing. The level of investment and resources committed are now a major financial concern.

Essentially, telecommunications strategic planning focuses attention on the services which could be marketed, in the case of public corporations, or in the case of telecommunications users, the services which the firms could require in the future. In both instances this requires an awareness of future technological options. Companies must assess the likely demand for future services and what might be charged for them against the capital and recurrent costs. Users' telecommunications strategic plans must be consistent with the wider strategic plans of their firms, and increasingly, as telecommunications technology advances, may involve reviewing the organization, management, and indeed raison d'être of the firm. For example, a firm relying on face-to-face contacts to trade must now face the trend toward electronic remote transaction of business. Insurance brokers may no longer be prepared to meet underwriters to arrange cover, and deal only with those prepared to quote premiums and underwrite risks over data links.

Telecommunications strategic planning also focuses attention on the systems required and other resources, including personnel. It is concerned not only with the procurement of hardware and software but also with the recruitment, training, and deployment of the staff needed in future. Increasingly, attention has to be given to the retraining of existing staff and in some instances the measures to cope with future redundancy. It may be necessary to develop an industrial relations plan to minimize the risk of future disputes. Implicit in the strategic planning must be consideration of the standards of service which the equipment and other resources are to satisfy.

Finally the telecommunications strategic plan must strike a financial balance between the future costs likely to be incurred against the income from users in the case of public telecommunications corporations, or against the cost benefits in the case of the private firm.

It would be quite unrealistic to expect such strategic planning directed at forecasting needs five years or more ahead to be precise. Greater precision is achieved as the strategic plan moves forward into the development and procurement planning stage. It provides a basis, however, for the later more detailed planning. Throughout the process the assumptions being made should be stated, and be subjected to analysis. For example the effects on the standard of service should be assessed for maintaining the same level of investment in the face of a higher rate of growth. It could be found that the consequences would be so serious as to justify a measure of insurance in the form of additional investment in spare equipment.

25

DEVELOPMENT AND PROCUREMENT PLANNING

Contracts are placed and other resources acquired based on development and procurement planning, and a greater degree of precision is required, therefore, than for strategic planning. Development and procurement planning is discussed under the following headings: determination of current usage of communications; forecasts of future requirements; identification of technical options; identification of external factors impinging on telecommunications planning; financial appraisal of options; preparation of a telecommunications development and procurement plan; and sensitivity analysis of a development and procurement plan.

DETERMINATION OF CURRENT USAGE OF COMMUNICATIONS

A first essential stage in any advance planning is a systematic assessment of the present situation, and telecommunications planning must start with determining what communications facilities exist and what use is made of them. Reliance can seldom be placed on records of what terminals exist and where they are. Reliable data on the level and incidence of usage is also seldom available.

There is often no alternative but to inspect and record on the ground what terminals exist and the facilities they provide, what line equipment exists, what switching capacity exists and what facilities it serves, and what exists in the form of a main network or outlets to other networks. A particularly close watch must be kept for terminals and equipment which exist unrecorded and possibly unused, the existence of which may not have been known. With the rapid growth in telecommunications the records of what equipment exists, and where, may not have been documented as well as they should have been. Expedient measures might have been taken to provide facilities urgently needed, and the procurement and utilization of equipment not so tightly managed as it should have been.

Given accurate detailed information on what equipment and other facilities exist, the next task is to assess the level and incidence of current usage. In taking the measurements it must be noted that usage will vary with the time of the year, the day of the week, and the time of day. The objective is to identify the pattern of usage, as well as peak demands. The measurements should enable a picture to be built up of between what points connections are being set up, the number of connections being set up between them, and their duration. Such traffic recording will continue over a prolonged period to ensure that peaks in demands between particular points are not being missed. While the measurements are being taken a watch must be kept for distortion or suppression of the traffic flow due to overload. The load being measured between two points may be underrecorded because connections which were required could not be established due to all the needed equipment already being in use. The measurements should be as detailed as practicable and left unaggregated at this stage. They will be an essential component of the design data needed for planning and premature analysis could result in the loss of information needed at a later stage. Telemetering equipment is available to record much of the data required for communications planning, but in some instances there may be no alternative to taking measurements by direct observation.

The measurements should be divided into voice and nonvoice, and subdivided again where appropriate. For example, nonvoice usage might be subdivided into message and data traffic streams. The usage of other nontelecommunications means of communication should also be examined to identify possible needs that could be covered in the plan. Obviously, communications by post might be carried by a future message service, and there may be other less obvious opportunities. The amount of travelling by staff might be reduced by providing teleconference or facsimile services for example.

There are few shortcuts in the collection of the above basic data. Later when the discussion of ongoing system surveillance is considered, ways in which the amount of detail to be analyzed can be minimized will be considered. The more detailed and comprehensive data is essential, however, for the preparation of a telecommunications development and procurement plan. Once collected the detailed information should be held on a data bank for analysis and processing.

Summarizing, information should be collected about the current usage of all forms of communication and broken down into main categories. The following might be the categories for a private firm:

Telephone calls		– internal
		– external
Messages	– by teleprinter	– internal
		– external
	– by facsimile	– internal
		– external
	– by word processor	– internal
		– external
	– by hand	– internal
		– external
Data	– by data links	– internal
		– external
Other		– internal
		– external

For each of the above main communication streams, the points between which communications are set up and the terminal used at each end should be recorded, with the duration of the connections and when they occur indicated. Thus for each internal telephone call the type of telephone being used, the numbers called over a representative period, and the duration of the calls should be logged.

FORECASTS OF FUTURE REQUIREMENTS

The first stage in forecasting future communications requirements is to examine any historical data available to identify any underlying growth trends.

For example, if records have been maintained of past representative usage it may be possible, by plotting levels of usage against a time base, to see that over a sustained period the level of usage has been rising in a way that provides a sound basis for projection. In examining past trends, however, the possibility that these have been influenced by exceptional factors has to be borne in mind. For example, the firm may have been going through a period of rapid expansion which resulted in what was essentially a steep increase in usage over a relatively short period which did not continue. Conversely, it may have been going through a period when for various reasons the level of usage was restricted. Or, external factors may have distorted usage trends, for example some national industrial dispute may have led to a greater use of telecommunications for a period.

Having projected future usage on the basis of past trends, the possibility of corporate plans for the firm being implemented and thereby affecting usage has to be considered. If for example the firm plans to increase its labor force significantly, the effects have to be allowed for. Obviously the nature of the staffing increase has to be considered: If for example the increase is in workers on the shop floor the effects on usage will not be as great as an increase in the number of office staff. Changes in the management, organization, and possibly the nature of the enterprise could have a major effect on usage. Finally, the introduction of information technology in the widest sense and the development of more advanced methods of communicating have to be allowed for.

It is essential in forecasting future levels of usage that the assumptions on which the forecasts are based are clearly stated. In actuality they are not all likely to prove well founded, or to apply with the precision allowed for. When adjustments have to be made later it is essential that they be made against a known base and not arbitrarily.

The degree of precision with which forecasts must be made varies according to the period being covered and their application. If a forecast is being made for example of the number of exchange lines to be served by a telephone exchange ten years hence in order to procure switching equipment, the degree of precision will be lower than forecasting the number of installation engineers required next week to meet users' demands. In the former case, not all the hardware will be procured ten years in advance, but space will be left for racking on which the exchange equipment can be fitted as it is needed with a relatively short lead time. In the latter case, the staff of installation engineers cannot be quickly augmented and if forecast requirements are widely adrift, either the labor force will be underemployed or worse, engineers will not be available to meet users' requirements.

As the forecasts are made they are analyzed as appropriate for different purposes, some serving a number of needs. For example, forecasts of terminals are required not only for the procurement of terminals, but also to determine the design of the local network. For terminal procurement the numbers and types of terminals have to be specified. For local network design the type of terminal is less important than their location throughout the system, which will determine the cabling distribution. Forecasts should be made to allow as far as possible for the development and expansion of the system piece by piece. It is uneconomic to provide equipment any further in advance of its utilization than is needed for its procurement, but at the same time the risk of service interruptions while the system is being enlarged must be avoided. The forecasts should, therefore, not only give longer term needs, but give intermediate forecasts for the progressive development and augmentation of the system. Generally forecasts are required twenty years ahead for such purposes as acquiring buildings, for example; ten years ahead for underground ducting and manholes; five years ahead for major switching equipment; three years ahead for underground cabling; and one year ahead for main network channels. The above design periods are illustrative and give no more than an indication of the periods which forecasts are likely to be required for. Forecasts tend to move down a "funnel of precision." The further away they are for, the wider their confidence limits, but in general the further away the time for which they are projected, the less precision is needed. Forecasts, however, are "rolled forward" and as the time for which they are made approaches, their confidence limits narrow and they are at the same time required to have greater precision. The possible exception to that, as a general statement, is the occasional overreaction of forecasters to immediate influences leading to vacillation about a midpoint forecast, which in the event proves to have been as sound as originally anticipated.

IDENTIFICATION OF TECHNICAL OPTIONS

Given quantified and detailed forecasts of the telecommunications to be provided for, the next stage is to identify the ways in which the technology available or in prospect could be applied to meet those needs. This could lead to some reexamination of the requirements, particularly of the facilities which were initially assumed to be needed. It could be that the examination of the technical options available could identify other facilities which could be available as alternatives. Telecommunications technology is advancing so fast that companies or private users investing in systems could face a real dilemma in this respect. So much that has been promised has been found groundless or at best premature, and users' confidence has been undermined. A further problem is that options may be offered which although technically impressive are either ahead of users' ability to use them or have no sound lasting use. At this stage the options should only be identified and operationally evaluated: A final decision will depend on their later financial appraisal.

The identification of the technical options can follow a logical sequence concentrating first on the terminals. Given that the prescribed functional requirements can be met, their physical characteristics and the ease with which terminals can be operated by users should be compared. Accommodation costs are a major ongoing item of expenditure, and the space needed for terminals, and particularly the extent to which the terminals can be integrated into the work station, should be examined. Ease of use could significantly reduce personnel training costs and facilitate the introduction of the new system. The extent to which terminals include software and operate independently of some central processor could be an important consideration. The extent to which voice, message, and data communications can be met from a single integrated terminal will be of increasing importance. Finally, the aesthetic design of the terminals available will be a material consideration.

Local distribution methods will be another major consideration, particularly the extent to which voice and nonvoice services might be carried by the same distribution network. The various arrangements that allow for growth and for the piecemeal augmentation of the network, and that provide for flexibility and movement in the location of terminals, should be examined and compared. In addition to the conventional twisted-pair cable networks, the options might include local broad-band cable systems provided primarily for cable television but available for other telecommunications applications, and cellular and other local radio networks.

Possibly the switching technology and equipment available present the most difficult options to identify and evaluate. They may include analog or digital carrier switching, and varying degrees of centralized programming and control of use. The range of user facilities could be extensive and crucial considerations could be the extent to which costs are increased by their provision, whether they are built-in or real options, and whether they meet identifiable and real user requirements, overt or latent.

Main network technology is advancing rapidly and there are major options to evaluate, including conventional cable and carrier systems, line of sight radio systems, satellite communications, and optical fiber systems. In terms of transmission performance the operating characteristics might be designed to offer a range of operational options.

Throughout the process of identifying and evaluating the technical options, the technical standard of performance should be emphasized. System failure rates, provision for alternative paths in the event of a system failure or overload, and the speed of restoration could be critical factors in the final choice of system. Integration with older existing systems and the complexity of the interfacing are major considerations.

No more than an indication has been given of the questions which should be posed in seeking to identify the technical options. A full technical assessment is essential, given the substantial investment involved in procuring and utilizing a telecommunications system. At this stage, the objective is to present the facts as an input to the final investment decision.

IDENTIFICATION OF EXTERNAL FACTORS IMPINGING ON TELECOMMUNICATIONS PLANNING

In practice neither public corporations nor private firms procuring telecommunications systems are likely to have complete freedom of choice in selecting a telecommunications system. Increasingly telecommunications companies, whether publicly or privately owned, are being subject to regulation, a topic that will be discussed in a later chapter. Even with the growing freedom users have to decide for themselves what terminals to own, they do not have untrammelled freedom. There must still be a degree of restraint and prior approval of what can be connected to a common user network, to safeguard everyone's use of the system. There remain a number of other external factors which impinge upon telecommunications strategy.

It is sometimes argued that the introduction of more advanced forms of telecommunications is creating unemployment. For example, that with the increasing reliability of telecommunications systems the number of engineers required to commission and service the systems has been reduced. But at the same time those that are employed need to be more skilled and their importance is greater, because when their services are required the need for speed of restoration and repair is greater. Again, for example, it is argued that the introduction of an internal electronic message-carrying system reduces the need for internal messengers, and those messengers no longer needed are entitled to redundancy compensation. The introduction of new telecommunications technology is subject to the same industrial considerations as the introduction of other new technology. The author would argue that these are resolvable, and that in the longer term there should be increased opportunities for employment and more satisfying occupations. This is not self-evident to those immediately affected, however, and the preparation of a telecommunications development and procurement plan should recognize that there are industrial relations issues to be resolved and include provisions accordingly.

Telecommunications companies cannot impose new technology upon users without preparing them for it. Failure to inform users and gain their understanding and support could have chaotic results or at best lead to misuse of the new system. The private firm applying new telecommunications technology may be better able to prepare its own staff for the use of a new system, but must be equally solicitous about carrying its customers and clients with it. If its clients and customers are accustomed to dealing with it in a face-to-face way, they may, for example, resist being expected to place orders by telephone or some form of data link. For a long time telex services were not used outside the firms in which they were in operation. There was a strong preference for sending documents by post, reflecting a belief that this was more secure and legally binding. Again, the slow development of the use of videotext services over the telephone, like Prestel in the U.K. and similar services, was not because they were unsound technically, but because inadequate thought was given to potential problems that could discourage users of the new facilities. It was assumed for example that the lay user would act in a logical way to select the frame of information needed. In practice such a degree of logic is seldom manifest, and no reliance can be placed on users following detailed complex instructions to operate a telecommunications system.

FINANCIAL APPRAISAL OF OPTIONS

Assuming the various technical options that will meet requirements have been identified, the next stage is to appraise them financially in the same way that any major investment would be evaluated. The appraisal is complicated by the pace of technological change and by the relatively short period before any contemporary generation of technology becomes obsolete.

A major issue for the private purchaser of new telecommunications systems is whether to purchase outright or to lease systems. If the decision to purchase outright is made, then a relatively short life for much of the system should be assumed in figuring its depreciation; certainly a far shorter period than the unit's actual working life. It must be assumed that much of the system will be replaced before it becomes unserviceable. If the decision is made to lease, then it is probable that the supplier will front load the terms of the lease, recognizing that the system supplied may have a relatively short operational life. In contrast parts of the system can be assumed to have a long working life, and the local distribution network could well justify outright purchasing and its depreciation being spread accordingly.

A further major consideration will be the assumed rate of expansion of the system and the ability to provide for this by additional investment at intervals over a prolonged period. A system which is being procured to meet a projected relatively high rate of growth, but which is underutilized in the meantime, carries a heavy cost penalty.

Financial appraisal of the various technical options is based on conventional discounted cash flow. The main expenditures break down into the categories of capital expenditures servicing costs; depreciation; leasing costs; accommodation costs; power and heating costs; maintenance costs; enhancement, expansion, and redeployment-of-system costs.

PREPARATION OF A DEVELOPMENT AND PROCUREMENT PLAN

Given financial ranking of the technical options, the next stage is the preparation of the development and procurement plan itself. This should cover the following:

1. The communications requirements the plan is directed to meeting. They should be quantified in terms of number and types of terminals, the level of the main categories of usage, and the facilities which will be available. The requirements should be projected for each year up to five years from the date of commissioning the new system, and ten and twenty years ahead with diminishing degrees of precision.
2. The hardware and software being procured, again for each year for the first five years, and ten and twenty years ahead.

3. Annual capital and recurrent expenditure required over the same planning periods.
4. Forecasts of the staff required to manage, commission, and maintain the new system.
5. Forecasts of the number of staff to be trained in the use of the new system.
6. Forecasts of the number of staff who will be made redundant or have to be deployed on the introduction of the system.
7. The arrangements for commissioning and changing over to the new system.
8. The measures to be taken to ensure that the new system is fully utilized as designed, particularly those taken to interface effectively with customers and clients.
9. The standards of service to be met by the new system.
10. Contractual safeguards.

The development and procurement plan should be "rolled forward" annually and updated. The second year of the initial plan becomes the first year and so on. Once arrangements are made for the annual preparation of a development and procurement plan the updating becomes the opportunity to review its implementation. Adjustments can be made in the timing of procurement to allow for deviations from the forecast rates of growth, and to take advantage of technological advances. By the beginning of any planning year it is probable that some 80% of that plan will be committed, given the long lead time of much of telecommunications equipment. Succeeding years of that plan will be less committed, and by planning procurement in advance investment can be more effectively controlled. Alternatively, if necessary, measures can be taken to ensure that committed investment is still effectively utilized, for example by either stimulating usage or by introducing deterrents if the capacity of the system is being prematurely exhausted.

ANALYSIS OF A DEVELOPMENT AND PROCUREMENT PLAN

It is prudent to subject a development and procurement plan to analysis, since its implementation will be spread over a number of years during which the forecasts and assumptions on which it is based may prove inaccurate.

The effect of the level of usage being higher and lower than forecast and rates of growth being higher and lower should be assessed. Companies operating common user telecommunications networks will have based their procurement plans on forecast levels of earnings, and should assess the effects of lower utilization than planned for on the return on their investment. This could indicate the extent to which usage might have to be stimulated to take up the capacity they have invested in, and is the justification for a company having a contingency plan ready to market its products and services at such times.

The sensitivity of the plan to higher usage and rates of growth than planned should be assessed. If the forecast level of usage is reached earlier than planned, standards of service will be jeopardized and user complaint and criticism could become embarrassing. It would be prudent to identify how resilient the system would be to overload and the extent to which relief measures could be relied on. It is particularly important to assess the extent to which shorter lead-time resources are provided, and the measures which could be taken to augment them in advance of such overloads. For example, what arrangements could be made to increase engineering repair staff in the event of bad weather and damage to the external plant? To what extent could engineers normally employed on installation be redeployed on repair? If for any reason the demands on the operator services became unmanageable, could clerical staff be employed temporarily on the switchboards?

The demand for telecommunications, unlike manufactured goods, must be met as it arises or the standards of service rapidly decline. Conversely, if demand is below the carrying capacity of the system in the short term there is no effective corrective action which can be taken. Unlike manufactured goods, telecommunications capacity cannot be stockpiled and used to meet demands when they arise later. Thus the importance of analysis in assessing the overall performance of the system if forecasts and assumptions are not fulfilled.

The private firm investing in a telecommunications system does not have the same need to assess the effects of underutilization on the return on its investment. It is concerned, however, with the effects on the standards of service of levels of demand rising faster than forecast. It is equally concerned if the rate of growth is lower than forecast and it proves to have invested more than it should have.

Other analysis may be less quantified, and focus attention for example on the consequences of additional equipment arriving later than planned and the need for contingency measures. Particularly important is consideration of the measures necessary to safeguard operations in the event of a major system failure. The more reliable the equipment and the less the likelihood of a breakdown, the greater the need to have prepared in advance for such an eventuality.

26

PREPARATION OF TELECOMMUNICATIONS UTILIZATION PLANS

Development and procurement plans are essentially concerned with the longer term development of telecommunications systems, and procurement of resources which cannot be immediately acquired or disposed of. The task for communications operational management is to ensure that the acquired resources are effectively utilized for the purposes for which they were intended. A telecommunications utilization plan is required so that actual performance can be set against it and significant departures from intentions identified and, if necessary, corrective action taken.

A telecommunications utilization plan should cover the current year in detail and with precision, and the second year in less detail and with less precision. It should be closely correlated with the development and procurement plan, to the extent of repeating some of the basic quantities. It should include the following:

1. The number of terminals actually in use against the number procured.
2. The load being carried by circuits and switches against their capacity.
3. The actual standards of performance against the prescribed standards.

4. The number of staff employed against the number justified by the level of system usage.

5. The actual expenditures under various subcategories against those budgeted for.

6. Achieved income levels against those budgeted for, for companies that are vendors of common user systems.

Implementation of the utilization plan should be reviewed monthly, and the need for corrective action decided. The temptation for management to overreact has to be guarded against, as well as the temptation to change the plan in the course of the year and to bring it into line with actual results which again is bad practice that merely produces self-fulfilling prophecies.

For major systems, utilization plans may be prepared for each level of management and appropriate to the decision making exercised at that level. For example, the utilization plan for the operator services in a particular area will focus attention on the loading of the switchboards, standards of operating performance, and the expenditure actually incurred against that budgeted for the operator services. The utilization plan for the management of the switched services in that area will focus attention on the loading of the equipment, its operational performance, and the expenditure incurred in the switching units serving the area against that budgeted for it. Utilization plans for higher levels of management will cover wider areas of responsibility, but only in the detail appropriate to the level of decision making, and will be derived from those for the lower echelons of management.

V

TELECOMMUNICATIONS OPERATIONAL MANAGEMENT

27

RESPONSIBILITY

Telecommunications management can be considered under two headings; strategic or overall management, and operational management. Strategic management is the responsibility of board members and senior management. They decide the services to be provided, and the standards of service to be maintained. They are responsible for acquiring the systems, equipment, and other resources including personnel required. They are ultimately responsible for financial performance and finally accountable to shareholders, the State, or both for the overall functioning of the system. Board members and senior management, however, can do no more than set the objectives and ensure that the means to achieve those objectives exist. Achievement depends on operational decisions being made at a level as near to the user and the point of implementation as possible.

In a widely dispersed service industry like telecommunications, operational day-to-day decision making must be delegated. Some of the reasons for a public system not meeting the requirements of users are failure to distinguish between strategic management and operational management, and attempts by the top of the company or even by the State to control the latter too tightly. Senior management, for example, may ensure that the engineers, vehicles, and other resources needed for installation repair are available in a certain locality. The effective use of those resources from day to day depends on the judgment of the local management which is on the spot and in a position to judge how best to deploy them in response to a changing demand for repairs.

The appointment of a competent local management is the responsibility of higher management, but once appointed it must have the authority and discretion to apply its best judgment. Its performance should be monitored of course and if over a period performance is unsatisfactory, it will be for higher management to take corrective action. But that is a longer term decision.

This division of responsibilities may appear self-evident, but unsatisfactory performance can commonly be attributed to failure to accept the realities of telecommunications operational management. The satisfactory operation of a telecommunications system depends on the many parts of it, and the work of a large staff, being integrated and coordinated. That integration and coordination is a primary responsibility of board members and senior managers. They cannot, however, personally make the countless decisions needed day-to-day to make the whole system work. Attempts to do so at best can lessen motivation of staff and at worst result in wrong decisions being made.

The basis of telecommunications operational management should be the local unit utilization plans referred to in the previous chapter. These should be prepared annually by local management and once approved, become the criteria against which the performance of local management will be judged. Telecommunications operational management is discussed below under achievement of standards of service and resource management.

28

ACHIEVEMENT OF STANDARDS OF SERVICE

Meeting users' expectations must be a constant concern of telecommunications operational management as well as achieving required standards of service. If users perceive the service to be unsatisfactory their reactions will be immediate and more important, if their perception is well founded their confidence in the service and their reliance on it will be jeopardized. It takes many years of good standards of service to gain a reputation for reliability but only a moderate deterioration over a comparatively short period to lose it. The achievement and maintenance of standards of service which are not only operationally satisfactory, but which users perceive as satisfactory, requires close and constant monitoring of performance to detect any incipient deterioration and correction of problems before users become aware of them.

OPERATOR SERVICES

From the earliest beginnings of telephony the service provided by switchboard operators was monitored to ensure that users were not kept waiting and that their demands were properly handled. This is just as important today. The switchboard operator remains a critical point of contact with the public in the case of public corporations, or with customers in the case of the private firm.

On the old style switchboards with connections being made by plugs and jacks, an observer can record the time between a calling signal appearing on the switchboard and an operator plugging in to answer it. The recorded times over a period of, say, half an hour can then be collated and the average calculated. Providing the service is good and sufficient calls are observed, this somewhat crude measurement provides a reasonable indication of the amount of time which callers are being kept waiting for. If, however, the service is not good and callers are being delayed, the observer is prevented from recording sufficient delay times for the average to be representative. Moreover, it is probable that any calculated average will conceal a wide dispersion of timings with some calls being answered quickly and others being kept waiting a long time. Switchboard operators are trained to answer calling signals cyclically. An operator on taking up duty should answer the lowest left-hand calling signal on the board, and should next answer the first one to the right on the bottom row of jacks. When he or she comes to the end of the lowest row the first calling signal appearing in the second row from the left-hand end should be answered. Thus there is a progression of answering signals in order, moving up the jack field systematically. The operator should not go back to answer a calling signal in a lower row of jacks than the last one answered until the top right-hand corner is reached, followed by a return to the bottom left-hand corner to repeat the process anew. This procedure was devised to avoid the "unfortunate call," the call which because it appeared high up in the answering jack field went unanswered because the operators always took the lowest one to rest their arms. This was particularly important on large switchboards where calling signals might be repeated to appear before a number of operators, and the load was thus shared among them.

In practice the "unfortunate call" remained a problem and was a constant cause of complaint, indeed was more frequently the cause of complaint than a poor average time to answer calls. Supervisors in charge of the switchboards watched calling signals closely to identify those which were not being answered in turn and directed the operators to answer them next. It was frequently the case that the likelihood of an "unfortunate call" was greatest when the switchboard was less heavily loaded, partly because of the probability of all operators being engaged on calls taking a long time to establish, and partly because the supervisors in charge relaxed their vigilance. The inclusion of one single call which had gone unanswered for a long time during switchboard observations could inflate as much as a whole month's average, and mask a generally good time to answer.

Switchboard management is concerned with far more than the time taken to answer calling signals, important though that may be in influencing public opinion or the opinion of customers. The way the caller is answered and the efficiency with which his requirements are met are equally important. Operators are intensively trained in the art of handling callers, elucidating their needs first in a businesslike but still courteous manner. This requires a detailed knowledge of the system and the services available in the case of the public corporation, and of the firm's organization and business in the case of the private firm. The costs of training a switchboard operator in the public service are comparatively high and operators continue to receive "refresher" training at about monthly intervals. In the early days, great emphasis was placed on ensuring that operators only used standard phrases and spoke in a rather artificial way in the supposed interests of clarity. (It was interesting to hear the operators on leaving the switchboard reverting to their normal mode of speech among themselves.) Today operators are given more discretion, the emphasis being on establishing good relations with the caller, being helpful, but not wasting his or her own time. Possibly the most common cause of irritation to the caller after being kept waiting for an answer from a switchboard, and having passed on his requirements, is to be left hanging on with no indication of what is happening. Public service operators are trained to keep callers informed about what is happening and to return to them at not more than thirty-second intervals. This practice is not always followed by private firms' operators who are inclined to leave callers waiting for an extension to become available for what may seem to the caller an interminable time with no further advice. Public corporations monitor their operators' handling of calls to ensure that not only is a satisfactory response time maintained, but callers' requirements are properly met. This practice can be resented by the staff as "listening in" and a form of "snooping" if not properly applied. Ideally the observer should sit next to the operator and take the occasion of any lapses of performance as an opportunity for personal coaching. Remote listening, although possibly operationally more convenient as it allows a greater number of operators to be covered in a limited period, is more likely to be resented. It would be poor industrial-relations practice to rely on the evidence of such observations for disciplinary purposes in any but the most exceptional circumstances.

By the very nature of their occupation switchboard operators work in more regimented and closely supervised conditions than other telecommun-ications staff. Switchboard supervision requires thoughtful management to avoid creating what can be a "school-room" regime. A good standard of service, on the other hand, requires close coordination and a flexible response to changing situations by the switchroom management.

Over the years many ways of avoiding the "unfortunate call" have been tried. On switchboards answering emergency calls from the public for the police, ambulance, fire, or other public services, the calling signals also normally set off an audible alarm to draw special attention to the urgent call. Other ways of giving priority to special calls have been used, and ways of delaying the appearance of other calling signals until the longest-waiting call has been attended to have been tried. This indicates the extent to which telecommunications companies have been attempting to avoid criticism of delays being experienced by callers to public service switchboards.

The use of plugs and jacks to connect calls via the operator is outmoded, although many firms continue to use them. Most companies have changed to cordless-type switchboards on which the operator establishes connections by keys and dedicated switches. Calls to such switchboards are "queued." Callers receive a ringing tone, but are not connected to an operator until one is free and signals readyness to deal with another call to the queueing equipment. The longest waiting caller is then connected to an operator, and thus, providing callers continue to hold on when receiving the ringing tone, they are answered in order of precedence.

Cordless switchboards require close management. Some people argue that in contrast to cord-type switchboards, now all callers are subjected to delay. If queues are allowed to lengthen there is some justification for the criticism. The switchroom management can no longer see at a glance the development of a load on the exchange from the appearance of a growing number of unanswered calling signals, but must rely instead on meters or other visual indicators of the number of calls waiting in the queue. (The larger units may have a number of separate queues and the load may vary between them.) Switchroom management must be more alert to take action in the event of a queue growing too long, normally by bringing additional operators onto duty, but in extremis by reducing the number of places in the queue. The maximum queue length can be set by the switchroom management according to the number of operators available, and when all places are occupied by callers awaiting an answer any further callers receive a busy signal. It is argued that it is better to turn callers away than to keep them waiting while listening to a ringing tone for a long time.

Compared with the cord type board the operators on a cordless switch-board have to make very little physical movement to answer and connect calls. While the work is less physically demanding, some of the sense of urgency when the exchange is busy is lost. When a cord switchboard was busy it was immediately apparent to everyone in the switchroom. There is very little difference in the ambience of a cordless switchroom whether it is busy or slack. The switchroom management possibly has to motivate the staff in other ways when extra effort is needed. Cordless switchboards are normally equipped with facilities for constantly measuring the time callers are kept waiting for an answer and for obtaining other data about the work rate of the operators and the load carried by the switchboard.

As well as assisting callers who are encountering difficulty in setting up their own calls, and providing special services, the operators also provide general inquiry and directory inquiry services. These require particularly care-ful management. On the one hand callers must not gain the impression that the operators are being less than helpful, and on the other the costs of the service must be contained. Again, attention is given to ensuring that callers are not kept waiting for an answer, but at a time when there are many inquiries taking a long time to deal with it is difficult to ensure that an operator is available within seconds to answer a call to the service. The work can be repetitious and boring, but demands a great deal of skill by the operators, first to find out from the caller precisely what his needs are, and then to obtain the information from the large volume of reference material.

Local knowledge can play a major part in answering inquiries quickly. Inquiry management involves continuous attention to the nature of the in-quiries being received and the augmentation of the operators' sources of information to ensure that they not only contain what is frequently needed but they present it to the operator in the most effective way possible. This is particularly important in the case of directory inquiries. Special local lists are normally held at directory inquiry centers of numbers requested by callers in forms not recognizable in the directory. It is not unusual for inquiries to be received as imprecise as the following: "Can you give me the number of the pet shop in the High Street?" The caller may not appreciate that directory inquiry centers often serve a wide area with several towns and not just their immediate locality.

The directories and other paper records used by inquiry operators are being replaced by visual display terminals with on-line access to central data banks. Using information retrieval techniques the answers to inquiries can be found faster, and the number of times a "no trace" reply must be given to the caller reduced. The problem remains, however, of obtaining sufficiently precise information from the caller—and quickly—to process the inquiry quickly and reduce the load on the service. It appears inevitable also that the better the service given by inquiry centers the more inclined the public will be to come to them and not take the trouble to look elsewhere first, for example in the telephone directory, to obtain the information they require. While companies providing common user services may not welcome the burden of inquiries and the substantial costs involved, the private firm may have no option but to provide such services for their customers.

One further major aspect of the operator services provided by companies requires close attention by management. The operator often has to charge the caller. The charge can be relatively large, particularly in the case of international calls. The details of the call by the operator connecting it must be recorded meticulously and much of the training of operators concentrates on this aspect of their work.

In summary, management of the operator services is likely to be directed to achieving standards of service based on the following measurements:

1. Percentage of callers requesting assistance answered in five seconds.
2. Time taken to connect callers requesting assistance.
3. Percentage of calls handled correctly.
4. Percentage of calls on which assistance was requested, but the caller could not be connected.
5. Percentage of assistance calls on which the correct charges were recorded.
6. Percentage of inquiry calls answered in five seconds.
7. Time taken to answer inquiry calls.
8. Percentage of inquiry calls on which the correct information was given.
9. Percentage of inquiry calls on which the required information could not be provided.

(Special services call standards would be judged similarly on the basis of 1–5.)

DIALED SERVICES

The successful establishment of a dialed/keyed connection involves the correct and precise operation of many separate components to establish an end-to-end path through the system just for the duration of that particular connection. The whole process of setting up a connection may be completed in less than a second with individual components operating in less than a thousandth of a second. Frequently the connection passes through a number of management units, no one of which is solely responsible for ensuring that the end-to-end connection is established satisfactorily. It is possible that no other operation involves the integration of such dispersed management simultaneously. The achievement of the high standards of service which users have come to expect requires highly competent, painstaking management and unrelenting attention to detail. It is not only the technical complexity of the operation which is demanding but also the sheer logistical scale of the operation. In the United Kingdom on a typical day more than 50 million telephone calls alone are dialed successfully by users. It has to be recognized, however, that even as low as 0.001 percent failures is equivalent to some 50,000 callers failing to obtain the connection they require at the moment they require it. Moreover, if those failures are occurring over a limited period and are concentrated in a particular part of the system the degree of user frustration can be high.

For many years reliance was placed on routine maintenance of the equipment to achieve the required standards of service. At the start of every working day, throughout the system engineers tested the functioning of as much of the equipment as possible, making sure for example that all the electromagnetic switches would operate properly and would rotate to select a path to the next switch in the train of connections. Then at regular intervals every piece of equipment was overhauled, whether it was failing or not. The emphasis was on preventive maintenance, replacing worn parts and readjusting equipment before it was a cause of call failure.

It came to be recognized that, the costs of such meticulous maintenance aside, the results of this philosophy were not achieving the required standards. At best the equipment took time to settle down again after an overhaul, at worst the very maintenance itself was creating faults which had not previously existed. A different philosophy was adopted: The equipment was left undisturbed until it went faulty and was then given a major overhaul—no matter how long or how recent the last overhaul had been. Reliance was not placed, however, on the users alone finding the failures. To have done so could have been disastrous.

Understandably, users encountering difficulty are seldom prepared to spend time reporting it, but would rather try again and if successful take no further action. (It is for that reason the assistance operators are instructed to hold any faulty connections they find and report them to the engineers for attention.) As management changed to the new maintenance philosophy, the functional testing of the equipment intensified. Automatic testing equipment was provided which could be left running unattended and which would set up many hundreds of test calls spread over the switching network. If a faulty piece of equipment was found it would apply a "busy" condition to prevent users having access to it, and print a record of the failure for the attention of the maintenance engineer. The testing equipment was normally run through the night and at other slack periods in an attempt to find causes of failure before the load on the system again built up.

Most telecommunications corporations continue to apply this philosophy, but with the increase in long distance and international traffic a major problem remains. Many failures occur only when a particular combination of pieces of equipment and circuits is used. If a test call is made which detects a failure over one of these longer trains of connections, a decision has to be made whether to hold it until engineers in some distant center can be informed and localize the cause of the failure, or to release the train of connections and rely on the faulty equipment being found in some other way. There is no unequivocal answer to the question. Some argue that such failures are difficult to reproduce and in the interests of users, once found should be held and the cause positively identified. Others argue that they do not cause so much trouble as to justify keeping a long train of equipment and circuits out of service, possibly for some time, until the cooperation of distant engineers can be arranged.

Engineering tests were designed essentially to find faulty pieces of equipment and circuits, preferably before they caused call failures, which could then be taken out of service and repaired. The incidence and type of failure were analyzed in detail and used to decide which items of equipment were most fault prone and where maintenance resources should be concentrated.

Such statistics, however, were not acceptable as a measure of the actual service which users were experiencing and it was argued that separate observations were needed to measure the actual standard of the dialed service, and that any failures observed should not be reported to the engineers for attention, because that would make the results unrepresentative. Many tele-communications corporations have relaxed such attempts to avoid any possible distortion of measurement of the standard of the dialed service. Many today use artificial traffic-generating equipment which continuously generates test calls spread widely over the network. The results are recorded by a data capturing system and processed by computer to provide detailed information on the incidence of call failures, including, wherever possible, the identity of the equipment and lines which failed the test. Frequently the conditions imposed during the test call are more stringent than would be experienced in normal use, again in an attempt to locate incipient causes of call failures. The results are passed to the maintenance engineers as soon as possible after they are available, and remedial action is often initiated before the end of the period covered by the service observations.

The introduction of electronic switching, with the connection of calls being controlled by some form of central processor, in any case required a new approach to the measurement of service standards. In many instances if the central processor detects that an attempt to set up a connection has failed, it will automatically attempt to establish it over an alternative path, often so quickly that the caller is unaware that a failure has occurred. The equipment or circuit which failed on the first attempt is automatically "busied" out of service and the failure brought to the attention of the maintenance center by the processor in the form of a printout giving the identity of the failing equipment. Many key items of equipment in the electronic switching center are duplicated and if at any time the primary equipment fails, the second is automatically brought into use as a replacement. At the same time the failure is brought to the attention of the maintenance engineer.

The management of the dialed service has increasingly depended on an understanding and analysis of data recorded by the central processors or other forms of data-capturing equipment. The manager of an old electromagnetic exchange was able by visual inspection of the equipment to judge the standard of maintenance and the quality of the service which users were experiencing. Modern systems have few if any moving parts, hence system failures are more difficult to detect and reliance has to be placed on monitoring and other facilities built into the switching equipment itself to detect any incipient cause of failure.

The management of the dialed service is concerned with not only the incidence of system failures, but also the loading of the equipment and the incidence of calls failing due to inadequate capacity. Various facilities are provided to measure the number of calls actually failing due to all the equipment being in use, and to measure at intervals the load carried against the capacity available. In the event of an overload, relief measures may take time to apply. In some instances local action may be possible by moving under-utilized equipment from one part of the switching center to the overloaded section. Usually, however, relief will involve installing additional equipment or bringing additional circuits into use, and will take longer. Local management should ensure that no available equipment or circuits are being held out of service awaiting repair, unnecessarily reducing the capacity of the unit.

Possibly the most difficult task in achieving a good standard of dialed service that its management has to contend with is the maintenance of transmission quality. Switching failures have been considerably reduced with the growth of electronic and digital switching, but in the reception of the distant transmission noise and other interference remain difficult to eliminate. What constitutes a poor quality of transmission, particularly on voice connections, can be a matter of subjective opinion in the last analysis. If it is agreed that the end-to-end quality of a connection is unsatisfactory, identifying the cause and locating the source in an extensive network can be a problem.

Management of the dialed services, lastly, must ensure that the charges automatically recorded by the switching equipment against users are correct. Problems include both overcharging and undercharging, the latter representing a loss of revenue.

Management of the dialed services is likely to be directed to the achievement of standards of service based on the following measurements:

1. Percentage of calls established satisfactorily at the first attempt.
2. Percentage of calls failing at the first attempt due to system failure.
3. Percentage of calls failing at the first attempt due to available equipment being in use.
4. Percentage of calls failing at the first attempt due to the required number being in use.
5. Percentage of calls failing at the first attempt because of no reply from the distant number.
6. Percentage of calls abandoned prematurely by the caller or on which the caller failed to complete dialing.
7. Percentage of calls on which the charges were overregistered.
8. Percentage of calls on which the charges were underregistered.

9. Percentage of calls on which the standard of transmission was unsatisfactory.

INSTALLATION REPAIR SERVICE

System failures are unlikely to deny users the use of their installations completely, or to delay the establishment of connections for very long. In contrast, installation faults can prevent any use whatever being made of an installation and thus even more importance is given by telecommunications operational management to installations.

In contrast also to the dialed service, very little can be done to identify incipient faults before they affect users' service. Equipment is now used extensively to monitor the electrical condition of lines in the local distribution network. The equipment can be left running continuously and tests every line in turn for low insulation, high resistance, and any other electrical condition likely to cause an installation to fail. More stringent conditions are imposed during the test than would be needed to ensure that the installation remained in service. Any faults found are recorded on a printout for the attention of the repair staff. Apart from that the only preventive maintenance that can be done depends on a visual inspection of the external equipment and cable to detect damage or other weaknesses.

Possibly more than any other telecommunications staff, the repair engineers work with virtually no close supervision. It is just not practical for the repair service management to see that every repair is properly carried out and a good standard of workmanship is maintained on every repair. The consequences of poor workmanship, for example clearing a fault without working to ensure that it does not occur again, may not become apparent until some time afterwards.

Repair service managers rely constantly on a close analysis of the incidence and categorization of faults to achieve good standards of service. They look for example for an increasing number of underground cable faults in a particular locality as an indication that possibly the local distribution cable serving it is damaged and "going down." They watch closely for frequent faults on the same installation, indicating either that it should be completely overhauled, or that the work of the repair engineers who have been attending to faults on it has not been up to standard.

Repair service engineers must rely largely on users reporting faults on their installations. Users understandably are not inclined to spend time reporting faults unless there is a complete failure of service. Worn cords, noisy connections, and so on are often not reported and a deterioration in their service is suffered, users not recognizing that the noise on their installation is also affecting other users attempting to call them. When eventually their service fails completely, the amount of work that has to be done to bring their installation up to standard can be quite extensive and delay the repair of other service-affecting faults. Most telecommunications corporations concentrate on restoring service immediately and record the non–service-affecting work to be done later when there are no immediate repairs to carry out.

The achievement of a high standard of repair service depends heavily on the skill of the staff receiving reports of service failures in obtaining precise information from callers about the nature of the trouble they are having, and diagnosing correctly the cause. Wrong diagnosis can result in the clearance of the fault being delayed while the time of the repair staff is wasted looking for it in the wrong part of the system. A range of skills are needed for repair work and it is often more efficient to employ different engineers, for instance on underground cable repair and on users' installation repair. It is essential, however, that the repair center diagnostic staff decide correctly where the fault is most likely to be and do not send an underground cable repair man when an installation engineer is needed.

The training and subsequent coaching of repair service staff should be a major concern of telecommunications operational management. They can have a major influence on the users' perception of the service, not least because they are often frequently in personal contact with the public, entering their homes and office premises. They have to deal with users who are possibly disenchanted with the service they are receiving, and the extent to which the repair service staff respond effectively and courteously will decide how quickly user goodwill is recovered.

Management of the installation repair service is likely to be directed to the achievement of standards of service based on the following measurements:

1. Average time taken to answer reports of service failures.
2. Percentage of service failures repaired by the end of the next working day following the report of the failures.
3. Average number of faults per installation per annum categorized according to the nature of the fault, and in total.

4. Percentage of service failure reports on which no fault was found when tested.
5. Percentage of service failures reported on which the cause was not subsequently found.
6. Percentage of service failure reports on which further difficulties were reported by users after the clearance of the original failure.

PROVISION OF SERVICE

Telecommunications operational management places as much importance on quality of service provision as on repair of installations. Users' opinion of the service is likely to be as much influenced by the speed with which their orders for installations are met and the standard of the workmanship as by any other facet of the service.

Success in achieving a good standard of service starts with the initial reception of the order, whether that be face-to-face with the user in a sales office, by telephone, or less commonly, by post. The marketing of installations is becoming more competitive since the long-standing monopoly or quasi-monopoly of the telecommunications corporations has been taken away from them. Sales staff must not only be trained to elucidate the customer's requirements but increasingly to seek to meet those requirements with one of the products marketed by their own organization. Telecommunications management is having to pay far more attention than it has in the past to training its sales staff to secure orders in the face of competition, and to encourage the customer to order more than he had perhaps intended. In the past some telecommunications corporations, particularly those administered by the State, have given the impression that to market additional products beyond the customer's stated needs was undesirable. In part this was attributable to a desire to contain the expansion of the service and partly to lack of confidence in being able to fulfill additional orders, given existing pressure on scarce resources.

A major concern of management remains to ensure not only that a competitive range of products is offered, but that having secured orders, adequate supplies of the products will be available and delivery dates will not be delayed because the product users ordered is out of stock. Major telecommunications companies face a dilemma in that respect. Normally lower prices can be negotiated with product manufacturers by placing orders centrally rather than by local managers placing their own orders. Central procurement and distribution of products inescapably introduces delays in the movement of supplies from where they are made, to where they are brought into revenue-earning use. Moreover, central procurement departments are inclined to be less motivated to see that customers' orders are met without delay, and more inclined to delay the award of contracts and otherwise to bargain with manufacturers in order to secure a lower price. The operational manager's attempts to achieve a high standard of provision of service can be seriously jeopardized by failure of the product supply pipeline. There is no perfect solution. If local management is left to make its own procurement arrangements with the manufacturers, it is doubtful that supply prices would be as low as with centralized procurement.

The total stocks of products held at all the local provision centers can substantially exceed those held with centralized procurement, even allowing for lengthy supply lines. To maintain such stocks locally so as to ensure customers' orders are met promptly, whether with central procurement or with local procurement, increases operational indirect costs and has to be set against the importance attached to speed of provision. Local management in any situation must maintain a close oversight of product stock levels and the storing of products to ensure that available supplies are used in the most cost effective way. If it is apparent that certain products are running out of stock and replenishment is likely to be delayed, then the local sales staff must be directed to conserve the stocks available as far as possible by not stimulating orders for them, and by suggesting alternatives.

The rapid development of the product range of terminals, with products becoming obsolescent almost before they have been introduced, has also created problems of local provision for service management. Manufacturers, having secured approval and endorsement of their new products by a telecommunications company are prone to unilaterally mount their own promotion campaigns. Customers as a result begin to place orders with local sales offices before a supply of the product is in place and more important, before the local sales and installation staff have been trained in its use and technical features.

For many years telecommunications companies faced demands for installations which far exceeded their resources, and they were unable to meet them without delays. Orders were placed on waiting lists or otherwise queued in the order in which they were originated. Much of the concern of local telecommunications operational management was with administering priority schemes and avoiding criticism that orders were being met out of turn. The situation has improved immeasurably and most companies should now be able to avoid any delay in the provision of their basic services. It has proved a major task, however, for local management to accustom the sales and installation staff to working without a stockpile of orders awaiting completion, and to be confident that a continuing flow of work will be ensured.

Seldom is an order for the delivery of an installation so urgent that arrangements must be made to meet it straightaway. It would not be appropriate, therefore, to seek to achieve a speed of response comparable to that of the repair service. Nevertheless, users expect to have their orders met within a short time and to be given a definite time when the installation engineers will come. Most telecommunications operational managements seek to operate an appointment scheme, striving to set a definite date and time for installation with the user at the time the order is taken. Meeting such appointments is regarded as more important in influencing user opinion than earlier completion of orders.

This does present problems for management and requires very close control of the work flow and the deployment of the installation staff. It also requires very close liaison between the sales staff taking the orders and the installation control staff to ensure that realistic appointments are made and can be kept. The flow of orders is seldom smooth and if the installation staff is to be used most productively, allowance must be made for that in arranging appointments. Wherever possible a buffer of nonappointment type work is maintained that can be used to keep the installation staff employed when all appointments have been met for the day. The difficulties of achieving a good standard of service are not made easier by the wide range of installation work to be done. Some jobs may only take an hour or so while others may take days, and tie down several installation engineers for a long time.

Telecommunications operational management in charge of provision of service must also ensure that when needed, line equipment is available for use in the local distribution network. Identifying and allocating line equipment for new installations can take some time and if not properly coordinated, can delay provision or lead to appointments not being kept. (Failure to have service on the day promised can be even more frustrating to the user than having to wait longer for an appointment.) Wherever possible sufficient spare line equipment is maintained to enable appointments to be made with the confidence that provision will not be delayed. Most telecommunications companies now have introduced information systems to process installation orders. As the order is being taken the details are sent simultaneously over data links to the department allocating line equipment, to the stores to have the necessary terminal equipment ready for installation, to installation control for the assignment of installation staff, and so on. Previously the order tended to circulate sequentially through the various departments concerned in meeting it, with inevitable delay and uncertainty about when it could be met.

Management of the installation provision service is likely to be directed to the achievement of standards of service based on the following measurements:

1. Percentage of service orders met on the appointed date.
2. Percentage of service orders on which service-affecting faults were reported within a month of the completion of the order.
3. Average time from the date when service was requested to be available to the date when the installation was provided.

In view of the wide range of installation work no single set of measurements are adequate for operational management, and they are divided into various categories according to the complexity of the orders.

SUPPORT SERVICES

The most critical support service to which telecommunications operational management must give attention is possibly the billing service. Users are quick to challenge the accuracy of bills, particularly for call charges, when they have no details of the make-up of the bills and no means of verifying their accuracy, and they are always inclined to underestimate their use of the service. Telecommunications companies inevitably have a responsibility to users in general not to inflate the costs of their services by giving in on contested bills and foregoing the revenue. On the other hand, they have to avoid giving an impression of arbitrary and unsympathetic attention to billing inquiries.

Most companies have incorporated checks into their billing computer programs to identify any abnormality in the pattern of a user's bill that might just conceivably have been missed by the extensive safeguards against error through the billing process. Any bill that is a departure from the trend of the user's past bills is identified and its accuracy rigorously checked before it is dispatched.

Telecommunications operational management has to strike a balance between, on the one hand, the lengths its staff are instructed to go in an attempt to secure the user's goodwill and retain his confidence, and on the other hand, safeguarding revenue. User opinion surveys suggest that whatever lengths the company goes to, there will remain a significant number of users who will continue to doubt the accuracy of the billing service. Management of the billing service is likely to be directed to the achievement of standards of service based on the following measurements:

1. Percentage of accounts corrected as the result of special checks imposed after the completion of the billing process.
2. Number of complaints of inaccurate billing.
3. Total charges waived following the investigation of billing complaints.
4. Percentage of users expressing satisfaction with the accuracy of the billing process.

29

RESOURCE
MANAGEMENT

Telecommunications operational resource management is concerned with achieving the maximum productive utilization of those resources already acquired or which can be obtained in a relatively short time scale. In practice this means maximizing the utilization of all equipment and other fixed assets already procured, and ensuring that all skilled staff are employed to maximum efficiency. In the short term there are few resources that can be augmented other than unskilled and possibly semiskilled staff.

EQUIPMENT UTILIZATION

Telecommunications operational management is concerned to maximize the utilization of existing equipment, first, to ensure that standards of service are not jeopardized by erosion of system capacity. (Reference was made earlier to having to be careful not to lose capacity through equipment being left out of service awaiting repair longer than necessary.) Secondly, failure to make full use of existing equipment will advance the need for its augmentation and incur additional investment in the system before it would otherwise be required.

Attention has to be given to the routing of circuits in the local distribution and main networks in particular to ensure that excessively long and circuitous routings are not used. Apart from taking up more line equipment than needed, such routings often require amplification and other additional equipment to maintain the quality of the transmission.

To a limited extent operational management may be able to move equipment from one part of the system where it is providing excess capacity, to another where an overload is developing, and thereby defer additional investment.

A significant part of the operational costs of a telecommunications system is taken up in testing equipment, mechanical aids (for example portable generating equipment and machine tools) and vehicles. The essential task of extending and maintaining the system must not be delayed through lack of resources, and at the same time, the maximum use of these resources must be the aim. The utilization of vehicles is a major concern, and detailed operational research is frequently needed to determine the optimum location for vehicle repair centers and parking places when not in use. The same location, for similar reasons, is also the optimum location for field engineers to report when starting work, and to hold field stocks of terminal equipment and other stores. Such centers should be located as near to the operational center of the territory served as possible in order to minimize time spent in travelling to work locations and returning at the end of the day or to collect further stores.

The running costs of buildings and other accommodations are substantial. Apart from ensuring that all available space is used most effectively and thereby deferring further investment in additional accommodation, operational management attempts to minimize heating and lighting costs and building maintenance costs. Telecommunications companies have extensive property holdings and often buy land well in advance of needs to secure a site needed for future expansion. Estate management can be a major concern of operational management as a means of augmenting income and offsetting system running costs.

The concern of telecommunications operational management to maximize the use of assets should not be satisfied, however, at the expense of jeopardizing longer-term standards of service. It is unfortunately too easy to defer expanding the system and thereby reduce investment in the shorter term, while allowing an overload to develop. At first the effects may be imperceptible, but in time users will perceive the degradation. The standard of service is then likely to decline further while additional equipment is obtained. With the long lead times necessary for much of the equipment, degradation can become very serious before relief measures can become effective. The short-term economy in investment can prove to have been gained at far greater cost than the amount saved.

Similarly any deferral of replacement of worn or obsolescent equipment in order to reduce investment in the short term can result in a serious deterioration in the standard of service which could take years to correct. The only grounds for deferring the replacement of obsolescent equipment could be the prospect of taking advantage of more advanced equipment that is being developed. It might well then be acceptable to allow a controlled deterioration in the standard of service, to allow time for more advanced equipment to be procured. Even then it must be recognized that the first generations of more advanced equipment in practice may prove to need further development before its reliability is established. It is not always prudent to be the first to invest in newly developed systems, whatever the competitive pressures.

MANPOWER PRODUCTIVITY

It is in the use and deployment of manpower that telecommunications operational management has the most direct and immediate control over the utilization of resources. By setting the level of recruitment against attrition of staff it has an effective way of adjusting the number of staff employed to the needs of the workload. The period over which this has to be done, however, may sometimes be extended. Skilled telecommunications engineers can take as long as ten years to become fully trained, and clearly numbers cannot be increased quickly other than by attracting trained engineers away from other companies. The rate at which many companies have been able to expand and develop has been constrained as much by a shortage of skilled telecommunications engineers as by any other factor. The acquisition of skilled manpower must be planned far in advance.

A major reduction in staff numbers must be planned equally far in advance, particularly at a time of high unemployment. At such times the attrition of staff tends to decrease, with employees tending to hold on to the jobs they have in face of limited opportunities for alternative employment. Some companies are subject to statutory regulation, which imposes rather stringent conditions on making personnel redundant. They face quite heavy financial costs if they wish to terminate the employment of staff prematurely and at short notice.

For many years telecommunications companies expanded rapidly with their manpower requirements constantly rising. It is easier to achieve higher levels of productivity during such periods, albeit at some risk to standards of service. In the last decade or so, however, many companies have been able to take advantage of the advance of the technology to reduce their manpower. The expansion of direct dialing, for example, has reduced considerably the number of operators required. It is fortuitous that investment, procurement, and commissioning of new systems take a relatively long time and offer scope for reducing the number of staff gradually and with less serious consequences for the individual employee. It is a fact of life also, that once the reality of the situation becomes apparent many employees will seek alternative employment of their own volition (to such an extent that sometimes operational management has difficulty in maintaining the number of staff required right up to the time the new system is brought into use.) Again, productivity may appear to be improving with the unsolicited reduction in staff numbers, but may conceal a degradation in the standard of service.

Telecommunications operational management has also faced a major problem in the last decade or so in manpower utilization from the rapidity of the advance of the technology. The skills of many of its staff acquired over many years have become obsolescent, and eventually they have become technologically redundant. Major retraining programs were set up and are likely to continue to be needed for many years as the technology continues to advance. Fortuitously again, with the long lead time of the investment this could be foreseen and planned well in advance. That operational management could be seen to be preparing for technological change in this way was a major factor in maintaining the morale of the staff. This subject will be returned to later.

Initial training of staff, their continued development and coaching, and increasingly their retraining, is a major operational cost which can be equivalent to over 2% of income, or more than £330 per annum per employee on average in Great Britain.

The continued development, coaching, and retraining of staff can present particular problems in a service industry. It is not always possible to allow staff time from their normal duties for training, and additional staff have to be employed to cover their absences. Many of the staff are widely dispersed in relatively small numbers throughout the territory served or are working in the field on their own. Training often involves significant time taken up in traveling to and from the training centers. Nonclassroom techniques such as computer-aided home study and videotapes are being used as far as possible, but are no substitute for "hands-on experience" of new equipment and systems. Telecommunications operational management faces another unique problem in maintaining the skills and expertise of its staff. It is not operating with a completely new system. New technology is being introduced piecemeal, with the obsolescent and the most advanced working together and interfacing. Its staff must not only be trained in the new systems and equipment, but also be expert in the old. Moreover, they must be expert in the interfacing of the old and the new; sometimes the most complex aspect of the whole system.

Like operational management in any industry, telecommunications operational management must ensure that its staff produces the best quality of work which it is capable of producing. Various methods of work measurements are employed to monitor productivity, but possibly more than in any other industry, emphasis is placed on the coaching and training of staff to fulfill their potential rather than on direct supervision at the work place. The development of the potential of subordinate staff is the prime concern of enlightened telecommunications operational management. Staff who cannot be directly supervised for much of the time, and whose substandard work may not be immediately apparent, must be encouraged of their own volition to give of their best.

A prime concern of telecommunications operational management in maximizing the utilization of manpower remains the effective deployment of the staff and ensuring that they have all the resources they need for fully effective work. No matter how well-trained and how well-motivated, staff not matched to the needs of the work and with no tools, stores, or other resources to work with cannot be expected to be fully productive.

Measures of manpower productivity are numerous and varied. Possibly more important than the absolute value of any measurement of productivity is its movement over time, providing the basis of the measurement remains unchanged. For example the productivity of the repair staff in a center may be measured in terms of the hours they work against the number of terminals in use in the territory served by the center. If over time the ratio lessens, indicating that less staff time is being employed on repair compared with the number of terminals in service, it might be assumed that the productivity of the center is increasing. Such a measurement has the attraction of a high degree of precision. The number of terminals in service would be accurate, and the hours worked by the staff should be obtained accurately from their paysheets. Even such an apparently basic measurement, however, can be misleading if not fully understood.

The type of terminals in use vary widely; comparisons between one center and another are invalid unless the number and range of terminals are taken into account. One center may have higher nominal productivity than another because it has a higher proportion of less complex terminals with a lower fault incidence. A reduction in the ratio for an individual repair center over time may be attributable less to higher manpower productivity than to an increase in the proportion of new, more reliable terminals. More seriously, a reduction in the ratio might be the result of spending less time on preventive maintenance when repairing faults, leading in time to an increase in the incidence in faults, and delayed degradation in standards of service. The man-hours assumed to be employed at the center may be misleading. Provision is made, for example, against a proportion of the staff at any time being away from work ill, while additional staff are assigned as reserves to cover such absences. If the provision is higher than the sick rate merits and is not allowed for in the measurement of productivity, a higher performance will be indicated than was actually achieved. Productivity can similarly be overstated by deferring training and taking advantage of the training reserve provision to increase the man-hours by not bringing reserve staff into the calculation.

INDUSTRIAL RELATIONS

When the rate at which telecommunications companies have expanded on the one hand, and the pace and extent of technological change on the other are considered, it may be seen as surprising that there has been so little industrial dispute involving telecommunications companies. To some extent this is attributable to the intrinsic interest of the work and to a sense of satisfaction coming from serving the public, in spite of ill-founded criticism at times, particularly by the media.

In general, however, telecommunications companies have recognized that possibly the most valuable operational resources they have, certainly in the short term, are their employees. Telecommunications are the most "perishable" commodity produced. A telephone call does not exist either before it is made or afterwards; it cannot be stockpiled; it is essentially an immediate service. The ability of the companies to counter the immediate effects of industrial action or "strikes" is accordingly circumscribed, and the cynic may argue that this has been a factor in the more enlightened approach of telecommunications to industrial relations.

Telecommunications labor unions tend to be farsighted in pursuing the interests of their members. Much of the disagreement between management and unions has been less about pay and conditions of service, and more about the measures being taken by the management to expand and develop services faster. While the telecommunications companies had a monopoly of common user services, many of their employees remained with them throughout their working lives. There was no alternative employment in which they could use their skills and expertise. That is now changing and there could be greater movement in and out of the employment of the major telecommunications companies, which could in time have a significant effect on industrial relations. In-house telecommunications unions may find that their own power bases are not so secure as formerly.

The telecommunications service industry is one in which there has been a large measure of upward movement, with many top managers starting at the bottom. They accordingly have a closer insight into the concerns of the staff they supervise. By the very nature of the work, first-line managers are in daily close working contact with subordinate staff and are not only able to do their work, but on occasion "handle the tools" themselves. Where industrial relations have deteriorated, it is possibly due to the failure of the companies to rely on the line management to keep its employees informed and consulted about its intentions, relying instead too heavily on union officials.

Many of the companies are State-owned or heavily regulated by the State, and possibly have accordingly been more open in publishing their accounts and making known their strategic plans. Many have a long-standing tradition of joint consultation, and staff and unions have in general been kept fully informed about their own futures. The growth in telecommunications, until the last decade or so, encouraged the staff to feel their employment was secure. Indeed the attractions of a secure job and the requirement of the State to set an example in good industrial relations have made employment in State-run communications very attractive and recruitment standards have accordingly been high.

Against this has been a tendency at times for both management and unions in the large State-owned companies to centralize negotiations. It is argued that settlements have been impeded unnecessarily, when local operational management and local officials might sooner have reached an agreement to the benefit of the user.

FINANCIAL PERFORMANCE

Telecommunications operational management is as concerned as any other with financial performance, but it is only at the corporate level that a realistic balance can be struck between income and expenditure. Inherently in a common user system income earned in one place depends on expenditure elsewhere. A significant part of the expenditure for example incurred in a particular switching center is to maintain equipment needed to complete a path through the system for a connection originated outside the particular management unit and for which it will not be credited with income. Conversely, connections originated in one management unit for which it is credited with income may depend on expenditure elsewhere in the system. It is unrealistic to judge local financial performance on profitability. Even more important, it would be unsound to allocate resources on the basis of local profitability.

The financial performance of telecommunications operational management inescapably can only be judged on the basis of unit costs. The annual costs per Erlang carried at a switching center can provide a sound measure of the cost effectiveness of the center. Again, however, such measurements can be misleading if taken at their face value. Clearly the age of the equipment will have a major bearing on such unit costs; the unit costs of the older switching center will inevitably be higher than those of a more recently commissioned one, to the extent that the relative efficiency of the managements of the two units may be the reverse of the order of unit costs.

Sound telecommunications operational management should be aimed at the achievement and maintenance of stipulated standards of service, and the control of expenditure to keep it down to levels provided for in the annual local utilization plan. Within the period of the budget, operational management must have discretion to vary day-to-day expenditures according to the transient needs of the situation. It has also to be kept firmly in mind that telecommunications operations are ongoing and cannot be neatly divided into accounting periods ending at midnight on March 31 or some other accounting deadline. Exceeding the annual budget by 0.5% is like using up the allotted funds on March 29 instead of on March 31, and is hardly within the confidence limits of forecasting anticipated when budgets were agreed some twelve months or more earlier. A period of bad weather impeding fieldwork toward the end of the financial year could increase the rate of expenditure by more than an annual rate of 0.5%.

It is important in a large-scale logistical operation like telecommunications that the staff and line management have an understanding of the underlying financing and operational costs. They can too easily be unaware of the financial consequences of the multitude of day-to-day detailed operational decisions they are making. For example, service engineers driving with scant regard to the fuel consumption of vehicles can be wasting money which might otherwise be available for the additional manpower they might be requesting. A major concern of telecommunications management is to inculcate in staff and line management a commercial awareness, and acceptance of an underlying concept that "what is not earned cannot be spent," while at the same time sustaining their concern for standards of service.

ACCOUNTABILITY

Reference was made earlier to the need, in such a large-scale logistical service industry as telecommunications, for decision making to be at a level as near to implementation as possible. It was argued that local operational management should have the devolved authority to use its judgment to respond in the best way possible to the varying needs of local users. That does not relieve local operational management of looking to senior management and board members for its stewardship and being accountable to them.

The organization, management style, and procedures for exercising accountability are critical in the overall performance of a company. Many telecommunications companies are State-owned, possibly still administered as a department of government, or are heavily regulated by the State. There is a tendency for management in such companies to be centralized and local management to be run essentially by the rule book. It is generally the practice for extremely detailed information about every facet of day-to-day operational management to be forwarded to the central administration, and for even quite minor decisions to be made at the highest level. Such organization, management style, and procedures are out of keeping with the needs of a fast growing, technology-based service industry.

Telecommunications operational accountability should be based on the utilization plans referred to earlier. These should be prepared in the first instance for the smallest management units and agreed on with the next level of management. The basic unit utilization plans should then be progressively collated at each level of management, being verified for consistency with any overall strategic direction given by board members or senior management, and provisionally approved. The collating at each higher level should be broader and in less detail. For example the utilization plan for a switching center may set a maximum call failure rate for the coming year. At the next level of management the maximum failure rate will be in terms of a group of switching centers. At the highest level the failure rate will be in terms of the whole system and will be the criterion on which the senior manager responsible for the performance of the whole system will account to the board for his stewardship.

The budgeted expenditure provided for in local utilization plans will be detailed, but again as the budgets are analyzed for higher levels of management they should be less detailed, and at the highest levels, be in the broadest terms. Senior managers will be financially accountable only in the same broad terms.

Utilization plans that include standards of service, resource allocation, and financial budgets should be provisionally approved as they are collated at each level of management. Inevitably, when finally collated the result may not be consistent with the overall intentions for the company and some adjustments may be needed, but this should be no more than is essential to bring the local plans and budgets into line with the corporate objectives. Utilization plans for the coming year should ideally be approved and endorsed for implementation about three months before the end of the current year. During the processing of the utilization plans the opportunity can well be taken for a first formal review of the likely results for the current year.

Board members and senior managers cannot be expected to, nor should they, abrogate their responsibilities by not monitoring in some way progress towards the implementation of the utilization plans and adherence to budgets during the course of the year. The extent to which they ask for information for that purpose is a crucial factor in establishing the standard of local operational management. Constant requests for detailed information and explanation, and attempts to intervene in local management will not develop the strong, decisive local management essential in a widespread service industry. An apparent lack of interest in local performance on the other hand will encourage undisciplined management and will be equally unhealthy. A balance has to be struck.

The telecommunications service industry is one in which a multitude of measurements are taken at the operational level to monitor standards of service and the utilization of resources. Such data is often produced as a by-product of the functioning of the system, and with the advance of information technology has become easy to capture, retrieve, and process. Such data is primarily required for day-to-day decision making. There is a tendency, however, to circulate such information too widely, with an inevitable tendency for recipients to seek further information about matters which are outside their responsibility. The circulation of information not required by the recipients for decision making tends to undermine the devolution of authority and at the same time does not serve the needs of accountability.

Sound accountability procedures depend on the careful design of management information systems. The fact that information can be obtained and widely circulated does not necessarily mean that it should be. As a general principle, information should be circulated only to those who need it to make a decision. In terms of accountability that decision may be a direct operational one, e.g., to intensify maintenance of part of the system, or an indirect one to intervene and direct that a subordinate manager take certain action, e.g., that he should intensify maintenance of a certain part of the system.

In general information should only be circulated upwards in the same degree of detail as the utility plan for which the recipients are accountable. It would be prudent, however, to augment this with some form of reporting exceptional information. For example, at switching-center-management level information is needed about the percentage of call failures for the unit. At the next level of management it is the percentage of call failures in a group of switching centers that should regularly be reported. If, however, the percentage of failures is say 10% higher than the planned standard for any individual switching center in the group, that exceptional fact should be reported to the higher level of management. The second level manager may decide not to intervene; knowing the competence of the first line manager for the switching center, he may have confidence that even before he receives the report corrective action will have been taken. Such reporting of exceptional information should be followed up the line of management.

VI

FREEDOM OF
CHOICE

30

MONOPOLY OF TELECOMMUNICATIONS

For almost one hundred years telecommunications users have had very restricted freedom to choose how their needs should be met. Apart from entirely internal systems, they have had to rent or buy terminals from the telecommunications companies owning the common user network, whether owned by the State, or by a public or private corporation. Even internal wiring normally had to be provided by the same telecommunications company. The user had to take whatever services, whatever standards, and whatever charges the telecommunications companies decided were appropriate. Such monopolistic power, albeit regulated directly or indirectly by the State, allowed users negligible freedom of choice. Manufacture of telecommunications equipment and systems were similarly constricted. Manufacturers were obviously reluctant to allocate resources to research and development and the production of equipment and systems until they were certain the telecommunications companies were prepared to go into contract with them for manufacturing these supplies for their "captive" users.

Given the need for the technical and operational integrity of telecommunications systems, it can be argued with some justification that such restrictions on user freedom of choice were necessary. Moreover, in the earlier years industry, commerce, and society were not so dependent on telecommunications; surface mail and other forms of communication were adequate alternatives to the telephone. Further, until the last decade or so the explosive growth in the usage of telecommunications and the growing demand for nonvoice services created such pressure on scarce technical and operational resources that the introduction of competition in the telecommunications field would, it is claimed, have been counterproductive and would have jeopardized what, in the event, was achieved.

The telecommunications companies argue that they were and remain concerned to provide services to the highest viable standards commensurate with what the user was and is prepared and able to pay, and that they exercised and continue to exercise their monopoly powers to the highest degree of responsibility. Nevertheless as industry, commerce, and society have become increasingly dependent on telecommunications the pressure has intensified for users to be able to choose how their needs can best be met. The pressure has undoubtedly been accentuated by the companies' own success in meeting growth and in developing services to higher standards of performance, and in so doing raising public expectations beyond the level they were able to meet, at least in the short term. There are grounds for arguing that telecommunications also became a popular political issue for members of both governing and opposing parties to focus public attention on, and for them to use the telecommunications companies for political ends. The growth in the potential market for telecommunications, particularly nonvoice, also created commercial pressures for other interests to be allowed to enter the field and to be allowed to compete with existing telecommunications companies, manufacturers, and suppliers. Whatever the strength of the arguments the fact remained that with restricted freedom of choice for users and no competition, there was no comparative basis on which the claims of the telecommunication companies could be evaluated.

Freedom of user choice is now becoming possible, although not everywhere nor as fast as many would argue it should. This book is about the complexities of telecommunications direction, management, and operation. It is now for the telecommunications companies to show in the face of competition that they have coped and continue to cope with those complexities, providing and developing services better than any competitor entering the field is able to.

This book has attempted to show in some detail what is involved in providing, operating, and maintaining efficient and cost-effective telecommunications. Major users will now have to cope with these complexities for themselves; organizations offering alternative networks and services to those provided by the long-established companies will have to face up to the realities which these companies have come to terms with over the years; and manufacturers entering the telecommunications field will also have to recognize these complexities.

A disappointing feature of developments in the telecommunications field over the last decade has been the lack of awareness of the underlying complexities on the part of those debating its future. The history of the development of telecommunications in the United Kingdom and elsewhere has a marked bearing on the weaknesses of the industry and on its future. Failure at times to recognize that fact has led to poor analysis and inappropriate conclusions as to the causes of these weaknesses. In the next chapter the history of the background of telecommunications in the United Kingdom is examined, as it is similar in many respects to that of other Western countries and the lessons to be drawn are as applicable elsewhere.

31

REGULATION OF
TELECOMMUNICATIONS

HISTORICAL BACKGROUND

The first common user telephone systems were provided by private companies or local authorities to serve particular communities. For decades, however, postal and telegraph services in most countries had been operated by the State as government departments. The developing telephone systems were seen as competing with the postal and telegraph services, and were very early on subject to licensing and regulation by the State.

As the telephone systems continued to grow and develop it soon became apparent that a network connecting the various local services was needed, and national systems began to develop. In many countries it was not long before the private companies were nationalized and ownership of the local authority systems transferred to the government.

In the United Kingdom the first telephone exchange was opened at 36 Coleman Street in London when the Edison Telephone Company of London was established in 1879. In 1880 the Edison Telephone Company and the Telephone Company amalgamated to form the United Telephone Company. In 1881 the Post Office opened a telephone exchange itself in Swansea. Eight years later the United Telephone Company and its subsidiaries amalgamated to form the National Telephone Company. (A National Telephone Exchange building continues to exist today and houses the main telephone exchange serving the city of Edinburgh. The original manual exchange has been replaced by automatic equipment and recently has been modernized with the most advanced electronic switches.) The National Telephone Company's trunk service was taken over by the Post Office in 1896, and finally in 1912 all National Telephone Exchanges were taken over by the Post Office. The last exchange operated by a local authority, in Portsmouth, was taken over in 1913. The only independent exchange system from then onwards on the mainland of the United Kingdom was in Hull, where in 1914 the Corporation was exceptionally granted a license to operate a system.

Telecommunications in the United Kingdom were administered until 1969 as part of the Post Office, as a government department. It is interesting, however, to reflect that in 1928 all Post Office extra-European telegraph services were transferred to Cable and Wireless Limited. In 1950 the control of all Cable and Wireless Limited's overseas telegraph services were transferred back to the Post Office.

As a government department the Post Office, including its telecommunications services, were subject to the same degree of oversight by Parliament as any other government department. Members of Parliament had the right to question the Postmaster General in detail about any aspect of the service and took full advantage of the privilege. They not only questioned the Minister about the overall direction of the Post Office but also queried the service being provided for their individual constituents. This constant questioning of the day-to-day operational management of the service led inescapably to caution and concern never to give grounds for criticism which could be politically embarrassing to the Postmaster General of the day. It was the convention that Ministers not only were credited with the achievements of their departments, but took full personal responsibility for any mismanagement by their staffs and were expected to resign if the circumstances were publicly held to be serious enough.

The financial control of the Post Office by Parliament was no different from that of any other department, albeit after the Inland Revenue and Customs and Excise it was the largest source of government income. For all practical purposes it was the only department that traded commercially with the public. All Post Office income was transferred to the Treasury account, and periodically the Postmaster General asked Parliament for an allocation of funds—the Vote—to meet the day-to-day costs of running his department. All investment was subject to the approval of Parliament, and again periodically the Minister asked Parliament for authority to borrow—the Loan—from the

National Loans Fund. All Post Office expenditure was included in the Government Financial Estimates presented to Parliament for approval after rigorous examination by Treasury officials. The opportunity of a debate on Post Office Vote or Loan motions was taken by Members of Parliament to question in minute detail the availability and standard of services in their constituencies. Much of the time of the Post Office Directorate was taken up by this close Parliamentary scrutiny and their commercial freedom was severely circumscribed. Inevitably Parliament took a very close interest in Post Office charges and fully justified increases to offset inflation were usually approved too late, and set too low.

In 1961 the extent to which Post Office financing was subject to Parliamentary approval was slightly relaxed. The periodical approval by Parliament of Vote expenditure was abolished. The Post Office retained the income it earned, met its own day-to-day expenditures, and could reinvest any excess after paying interest on loans and providing for depreciation. It could not finance its investment entirely from its own resources, however, and the Postmaster General continued to have to go to Parliament for loans from the National Loans Fund. The close oversight of the day-to-day management of services continued, with the Directors of the Post Office periodically being called to appear before Committees of Parliament to be questioned directly on their stewardship.

The Post Office was not allowed to decide the rate of pay and other conditions of employment. Employees were graded as civil servants or integrated with civil service grades, and their recruitment was subject to the approval of the Civil Service Commissioners.

By 1971 telecommunications' annual income has risen to £786 million, expenditure including provision for depreciation and interest on outstanding capital to £692 million, and the net value of fixed assets to £2415 million. Return on capital, regarded as profit plus interest on capital plus supplementary depreciation, was 9.8%. In that year the number of exchange lines in service was increased by 7.8% to over 9 million. This was only equivalent, however, to 16.6 per 100 population and there was still far to go before the telephone would be as widely available as in other Western democracies. There were almost 15 million telephones in use equivalent to 26.8 per 100 population. There were over 9000 automatic exchanges in service, but over 97% were aging, obsolescent Strowger electromagnetic exchanges. There were still 111 manual exchanges. Modernization of the service was long overdue. There were nearly 14 million cable pairs in the local distribution networks and nearly

100,000 speech channels in the main network. During the year about 11,000 million telephone calls were made, and the number was growing at over 11% per annum. Over 230,000 staff were employed, of whom some 201,000 were in the areas providing, extending, and maintaining the telecommunications services in the field.

By 1969 it had become increasingly apparent that its administration as a government department was seriously hindering the growth and development of the Post Office as a major service industry. In 1969 the Post Office was made a public corporation: the ownership of the service was retained by the government, but day-to-day responsibility was vested in a board. The board was appointed by a minister; the chairman and a number of members were full-time with executive responsibilities. Other board members were part-time and came from a wide range of outside fields, including the trade unions. The minister had to be consulted on major issues such as tariff increases and retained the power to approve the annual investment plans of the Post Office. The board had little freedom to borrow other than through the National Loans Fund.

The corporation continued to have a monopoly over postal communications and telecommunications, but its freedom was regulated through Acts of Parliament. Members of Parliament could no longer question the responsible minister about the day-to-day running of the services, but the chairman and board members were called before Select Committees of Parliament from time to time to be interrogated in detail about their stewardship. Statutory Users National Councils were also established and the board was required to consult them on any major decisions affecting the services. The Users Councils also investigated complaints referred directly to them of mismanagement or unsatisfactory service.

The relationship between Parliament and a nationalized industry has been controversial since the earliest days, and relationships between the government and the Post Office Corporation were no exception.

The board was constantly criticized for failing to meet the expectations of users. It faced, however, a number of intractable problems. It was responsible for both postal and telecommunications services, and at best this created conflicts of interest within the board. As in many other countries the postal services were running at a loss and were subsidized by the telecommunications side of the business. Decisions that might well have been in the interest of the telecommunications segment could have reacted against the interests of the postal segment. Time was needed to disentangle more than half a century of completely integrated management, and to develop policies that would enable both businesses to pursue independent freedom of action. The government could not allow the board immediate and complete freedom to charge and spend as it wished; it had too large an influence on the economy. For example, the pay of telecommunications staff was regarded as establishing the norm for annual pay increases, both in the postal services and in the wider economy.

Of possibly greater significance from the board's point of view was the tendency of governments of both political parties to use the Post Office as a regulator of the economy. It was not unusual for the level of investment to be cut at short notice on the grounds that the economy was overheating. On occasions this led to applications for service being stockpiled although equipment was available, because the installation staff had been reduced. Installation staff pay was regarded as capital expenditure on the grounds that it was adding to the value of the asset base and that it should accordingly be amortized. Worse, when investment in line equipment and exchanges was cut, with the long lead-time of the investment the consequences continued long afterwards in the form of waiting lists for service. It might be argued that had the Post Office been allowed to maintain the level of investment which it had already provided for in its financing, it would have taken more money out of circulation in the form of connection charges than cutting investment would have. Whatever the force of this argument, the fact remained that possibly because the Post Office was once a government department, telecommunications expenditure was still regarded as included in the public sector finances. Government, attempting to constrain public sector borrowing, still treated British Telecommunications as a quasi-governmental department. In the late 1970s and 1980s the corporation's commercial freedom was further restricted, and just at a time when it was facing competition from the private sector.

The Post Office was criticized on all sides for not meeting users' expectations, and was charged with mismanagement. British Telecommunications manufacturers were understandably discontented with a situation in which their major customer constantly varied its level of purchasing, making their production planning extremely difficult.

In spite of all the difficulties, by 1981 telecommunications' annual income had risen to £4554 million; expenditure on outstanding capital, including provision for depreciation and interest, to £5250 million; and the net value of fixed assets was £14,614 million. In that year the number of exchange lines in service was increased by 4.7% to over 18 million, an amount equivalent to 33.1 per 100 population. There were nearly 28 million telephones in service, equivalent to 49.8 per 100 population.

By 1981 there were well over 18,000 exchanges in service. More than 12% were advanced electronic exchanges and a significant start had been made on system modernization; the last manual exchange had been replaced by an automatic exchange in 1976. There were over 26 million cable pairs in the local distribution networks and more than 240,000 speech channels in the main network. Over 20 billion telephone calls were made during the year and the rate of growth continued at well over 5%. By 1981 the number of staff employed in the field had been reduced by 4% to 193,000.

In the ten years after the Post Office was made a public corporation the telecommunications services grew as much as they had in the previous fifty years. Whether the board could have done a better job is a matter of opinion; it decidedly could have done a worse one. For much of the time it faced an explosive growth in the demand for its services while faced with inadequate resources; and inevitably standards of service suffered. Some argue that the board should have safeguarded its existing users' service, even if necessary by keeping new applications for service waiting longer. In the event, it tried to strike a balance and inescapably was criticized by both existing and potential users. The morale of the field staff had constantly to be strengthened in the face of what was seen by them as unjustified and at times unanswered criticism. The board was in a dilemma when faced with criticism by the media. To offer any rejoinder merely stimulated further complaint and provided ammunition for certain journalists who made a personal campaign of attacking the Post Office.

Critics of the Post Office and other telecommunications companies consistently show little understanding of the logistical scale of the enterprise, or that the margins deciding success or failure are extremely small. For example, with the number of telephones in service growing by more than one million a year, an underestimate of only 0.25% in the annual demand can mean the completion of some 25,000 orders being delayed and placed on the infamous waiting list. The boards and management of major users, of companies offering services in competition with the long-established telecommunications corporations, and of manufacturers entering the field will find the degree of precision required in decision making is unexpectedly high. The basic statistics for the British Post Office given in the Appendices illustrate the scale of telecommunications logistics and financing.

LIBERALIZATION

In 1981 separate boards were appointed by the United Kingdom government for telecommunications and the postal services, and the links between the two businesses were severed. Of greater significance was the legislation passed to counter the natural monopoly of the still state-owned corporation and the freedom granted for private suppliers to compete with British Telecommunications.

Users no longer had to obtain terminal equipment, including telephones, from British Telecom, but any terminals connected to the BT network had to be approved. Approval of equipment was still essential to safeguard the technical and operational integrity of the system. Approval was no longer the responsibility of British Telecom, but was transferred to an independent approvals board. Contrary to an opinion generally held by suppliers, BT had long been somewhat embarrassed by its authority to approve terminal equipment and subsequently being charged with obstruction and delaying private development when it had to withhold approval, or require further development or modifications. The approvals board was soon being criticized in turn for delaying prototype approvals, faced as it was with a flood of applications.

It is argued that the creation of competition in the supply of terminals forced BT to step up its marketing and make greater efforts to meet users' demands. There may be some force to the argument, but another factor was the decline in unsolicited orders for service since by the early 1980s the top of the exponential curve in the growth of telephony was reached. In many parts of the country more than four out of five households had the telephone: by 1982 the annual growth rate in exchange connections was down to 3% and that of telephones in service to 2.1%. The waiting list for exchange line service had been virtually eliminated and throughout the system margins of underutilized equipment existed which at long last justified promotion of the telecommunications market. 12% of the total exchange capacity and 24% of all lines in the distribution networks were unused. For whatever reason BT was fortunate in having the resources to compete with that it had not had previously. In some measure the progressive improvement in the financing of the corporation through the 1970s was attributable to the high rate of growth. The connection charges paid by new users was "new" money, and there was, therefore, another reason for BT to compete strongly and encourage the growth of new business.

The government was, understandably, watchful, however, that with British Telecom still having a monopoly of the distribution, switching, and transmission of telecommunications, it should not knowingly or otherwise favor its own terminal marketing. British Telecom was open to the charge of giving precedence to the connection of its own terminals over those supplied to users by its competitors. The government directed, therefore, that within BT the marketing of terminals should be clearly separated from the management of the system, and that separate accounts be prepared for the two. In practice such a separation has been extremely difficult to organize. In the field the processing of orders for service, installation, after-sales servicing, and charging were operationally fully integrated. To establish a separate sales force for the marketing and supply of British Telecom terminals would take time and possibly be less cost-effective than a single sales force. As an interim arrangement, internal transfer accounting arrangements have been made to create a nominal separate profit center for BT terminal marketing.

The care taken to ensure that BT does not favor its own terminal marketing possibly overlooks a counter consideration. It will be argued later that the real problem for British Telecommunications is to increase the loading of the system and increase telecommunications usage. The view might be taken, therefore, that in time, if not immediately, BT will welcome the marketing of terminals by competitors as additional sources of system usage. Given the ephemeral nature of much of the terminal market where products rapidly become obsolescent, it could be that BT in time will give less priority to marketing terminals and more to marketing telecommunications systems and usage. The concern not to divide British Telecom into a number of separate companies, each concentrating on a particular aspect of telecommunications, although more attractive to the telecommunications unions, may in time be seen as commercially less advantageous. The unions clearly would not like the labor force divided between a number of companies with the potential erosion of their negotiating power. On the other hand, it might result in more dynamic commercial management if responsibility were so divided.

While British Telecommunications had a virtual monopoly on the supply of terminals for connection to its system, it was under a *de facto* obligation to meet users' demands. It had no freedom to be selective about markets, whatever the commercial advantages or disadvantages. It was also inhibited from entering markets which did not arise directly from the provision of public telecommunications. Now it is probable that in time BT will decide that certain terminals do not justify marketing and are best left to competitors.

Already it is entering markets which it would previously have considered to be outside its purview. It clearly sees opportunities for doing so in the non-voice field and in software sales. Yet in the past, while it was failing to meet demands for its basic services, it would clearly have been impolitic to divert resources to entering other markets and competing with private suppliers.

The installation of terminals is labor intensive and is difficult both to organize cost effectively and to achieve a good standard of service. Again in time BT may well decide that installation is best left to competitors, particularly the smaller jobs ideally done by local small contractors or even by the user himself. The telecommunications unions may object to the loss of the work, but a comparison with the installation of electrical appliances makes the point obvious.

COMPETING NETWORKS

At the same time it allowed competition in the supply of telecommunications terminals, the government licensed Mercury, a private company, to establish a national network separate from BT and to compete with BT for the transmission of telecommunications. In addition to the arguments in favor of providing the stimulus of competition to BT in the transmission of telecommunications, advantages were seen in having an alternative network for use in the event either became unavailable for whatever reason. The establishment of the Mercury network, and its interfacing with the BT system, has proved far more difficult than introducing competition into the supply of terminals. The implications are also more far-reaching.

The capital costs of setting up an alternative network are very large and a commercial return on the investment would take a long time to achieve. Nevertheless, if Mercury could secure a relatively small percentage of longer distance telecommunications income its commercial future would be secured, particularly bearing in mind the continued high rate of growth of usage. The task of designing, procuring, installing, and maintaining an alternative national network to that built up by BT over decades is formidable. Mercury faced two other major problems as well, in achieving its objectives.

The number of large users of telecommunications who would be prepared to rent even dedicated or "exclusive" circuits from Mercury was limited and reliance could not be placed on them to generate sufficient traffic to load the Mercury system adequately. To set up an alternative common user network to BT involved reaching agreement with BT on the interworking of the two networks. Not all of the traffic originated by those served by the Mercury network would necessarily be destined for other Mercury users. Some traffic could only be finally carried to its destination over the BT network, and BT would expect a share of the income. Some BT originated traffic, conversely, could be destined for users by the Mercury system, and Mercury would expect a share of the charges collected by BT.

The second and possibly even more difficult problem that Mercury faced was its dependence on BT's local distribution network to connect many of its potential users to its network. Some it might reach by line of sight radio links, but the costs would be high, and for the majority the BT local network was the only viable option. BT again understandably was not prepared to make it easy for its competitor and negotiations on both counts were prolonged and difficult. The two sides reached an agreement only after the government intervened.

It will take time for the Mercury network to become viable and for the agreements which have been reached to meet the aims of government, British Telecommunications, and Mercury. To allow time for Mercury to be successful, the government has announced its intention of not licensing any further national networks until the next decade. It is possible that a solution to Mercury's problem of connecting its users to its network might be the local cable networks licensed for cable television. The cable television companies need to find alternative uses for their networks to earn a commercial return on the heavy investment they are facing. Such networks could potentially be used to connect Mercury users to its national network, and the cable companies could be paid an access charge.

CELLULAR RADIO NETWORKS

A further significant counter to the monopoly of British Telecommunications was created by the licensing of two cellular radio networks, one to be operated by British Telecom in partnership with a security firm, and the other by a private partnership. It is too early for the full potential of this development to be apparent. When the costs of the mobile terminals have been brought down as they inevitably will be, and the systems are shown to be operationally reliable, cellular radio will be a very competitive alternative to either the existing networks or the cable television systems for the local distribution of telecommunications.

The cellular radio network licenses which have been granted stipulate that the two networks should interface and be technically compatible. Thus users of one service will still be able to communicate with users of the other, and with the BT and Mercury networks. This is essential if the use of the systems is not to be severely limited, and a truly common user service is to be created. Again, agreement on the internetwork charging arrangements was not easy to negotiate.

VALUE ADDED NETWORK SERVICES

Yet a further counter to the British Telecommunications monopoly was created by a requirement that BT allow the use of its network for value added services. This regulation requires BT to hire out its network for the transmission of telecommunications services which it does not currently provide and is not planning to supply in the future. The renter of the network who does provide such services is allowed to charge more than BT's charges to reflect the value of the added service.

Traditionally the Post Office (and subsequently BT) had only permitted the use of its network for communications via its common user network or over dedicated circuits to a distance terminal of the same user. The objective was to force as much traffic as possible onto the common user network, where it could be metered and charged for accordingly. The charges for dedicated circuits—"private wires"—were set relatively high to compensate for the loss of metered revenue. Such restricted use of private wires was relaxed, in time, to allow their use for carrying communications between terminals of two different users, known as "A to B" usage. Again, the charges were set relatively high to offset the loss of metered revenue, and to pay for using the common network as a standby alternate for overflow conditions or system failures. The Post Office, and later, BT firmly resisted pressure to allow the use of private wires for the transmission of telecommunications to a third party, "A to B to C" usage. For example a user in London with a private wire to another user in Birmingham could not set up a connection to a third party in Birmingham via the local switched network. If this had been allowed, renters of private wires would have been able to bypass the main network and get trunk calls, or long-distance calls, for the price of a private wire plus a local call. BT clearly sees value added service as a further erosion of the return it should be getting for the heavy investment made in its network.

The full implications of the regulation are not yet apparent. Those groups pressing for the freedom to provide value added services argue that if BT is not prepared to meet the demands of users, it should not prevent others who can do so in a commercially profitable manner from doing so.

Probably the potential areas of expansion for value added services lie in the nonvoice, or data, field, and in developing nontraditional services. Firms renting out office accommodation, for example, may see an advantage in providing common user information-service data banks over the BT network and charging tenants for the facilities. They might even feel it worthwhile to provide tenants with telephone facilities which they rent in a block from BT, and relieve tenants of the burden of arranging their own telecommunications. Such arrangements are common for heating, lighting, and other standard services, and there are no apparent reasons why these arrangements should not be extended to telecommunications facilities.

THE FINANCIAL IMPLICATIONS OF LIBERALIZATION

In any major common user telecommunications network it is extremely difficult to break down expenditure and income to determine the profitability of various sectors of the business, such as one territorial division against another. Whatever accounting conventions are used, the information obtained from the field about the deployment of manpower, equipment, and other resources between different activities is not very precise. For example, a maintenance engineer in the course of a normal day may work some of the time on nonvoice services and some of the time on voice services or other duties. The engineer's records of the allocation of this time can only be taken as an approximation. The more detailed the breakdown asked for, the more doubtful its validity.

Even more doubtful are attempts to allocate equipment costs to various services on any basis of relative utilization. For example, line equipment in the local distribution network may be used most of the time for local telephone calls, but the same equipment is used for long distance telephone calls and occasionally for international calls. The apportionment of costs of the local distribution network among these three services can only be an approximation. Any attempt to assess the profitability of long distance or international calls without taking local distribution costs into account, however, would clearly be illogical, since long distance and international connections could not be completed without the local network.

In spite of the accounting difficulties, every company must assess the relative importance of its individual services and the variances between territorial costs in order to determine the charges it should make. When a company has a monopoly of communications it can set its rates to achieve an overall rate of return, accepting that some services may yield a low rate of return or in some localities may even be provided at a loss. In other words, some services are cross-subsidized and in some localities customers are charged less for services than elsewhere. When that company faces competition, its freedom to offset high costs by charging more is constrained. It may be forced to raise its charges for services earning a low rate of return or in localities where its costs are not adequately covered, and lower its charges for services earning a high rate of return to counter competition.

In the past Post Office, and subsequently British Telecommunications, charges were set to achieve an acceptable overall rate of return, the increases in charges frequently being limited to what the government considered were in the national interest, and what public opinion would tolerate. While the system was entirely state-owned, such decisions were understandably made on political rather than commercial grounds, and while under political domination, it was difficult to resist the pressure of public opinion. The result was that inevitably a large measure of cross-subsidization between services and between different parts of the country developed.

It is often the case that 80% of the users of a public telecommunications system generate as little as 20% of the revenue, but account for considerably more of the capital costs of the system. The heavy users are subsidizing the service of the remaining majority of users. Users value long-distance services relatively more than local services, and the markets for international, and to a lesser extent domestic, long-distance telecommunications will bear significantly higher charges than local ones. For the reasons discussed earlier cost comparisons are imprecise, but even taking the most unfavorable assumptions, international and domestic long-distance services are very profitable.

It is possible that the profitability of the long-distance service is not fully revealed in companies' accounts. In the past the cost of long-distance transmission of telecommunications was closely related to the costs of the network connecting the switching centers. Now network costs have been substantially reduced in proportion to the volume of traffic carried, and switching costs are relatively higher. The number of stages of switching on a local connection may be significantly greater than on a long-distance connection, and local connection costs accordingly relatively higher.

Whatever the objective basis for cost apportionment, comparisons also depend on value judgment. For example it is argued that the costs of providing a public phone booth service are greater than the income they bring in. It might be argued that the public phone booth provides the user with a means of calling his office or home when he is away and is part of the overall service provided by a telecommunications company: Few calls made from public phone booths are to other pay phones; most are to business or domestic terminals. Users given a clear choice may be prepared to accept a marginal increase in their regular charges in order to be able to receive calls made from public places. A measure of cross-subsidization might well be acceptable in the matter of phone booths, providing its extent is known and a decision is consciously made to allow it.

It is argued that directory inquiries should be charged for in proportion to how often they are used. The arguments pro and con are evenly balanced, but there is a case for regarding directory inquiries as part of the overall service provided by a telecommunications company, and that the costs should continue to be met by marginal increases in users' regular charges.

Whatever the views held, cross-subsidization of telecommunications services existed when the monopoly of British Telecommunications was reformed and competitors were authorized. Clearly there is a real risk that competitors acting commercially will seek to take the most profitable business from BT. If it is to maintain a proper rate of return, and more importantly to sustain the standard of its services, BT might have no alternative but to reduce the prices of its more profitable services in order to defend its market share. At the same time, to offset any loss of income, it would have to disproportionately increase the charges for its less profitable services. It might in time feel unable to continue unprofitable services, and reduce its investment in those parts of the country where costs are relatively higher in relation to earnings. This possibility was foreseen by the government, and BT's operating license places an obligation on the corporation to maintain services when it might not consider them cost effective from a commercial standpoint. The government has also limited the price increases which BT can make in the near future, and has established a regulatory authority to ensure that its intentions are followed.

Some believe that British Telecommunications could "unpackage" its charges and eliminate cross-subsidization without dramatic increase in its charges because its overall financial performance is strong. Others argue that it could change its charge structure without major price adjustments by raising the productivity of its staff. Any large organization should always be looking for ways of increasing manpower productivity, particularly when it is investing heavily in new technology. The comparisons drawn earlier between the years 1971 and 1981, which saw as much expansion in ten years as the previous fifty accompanied by a 4% reduction in field staff, show the progress made.

Undoubtedly more progress will be made, but it should not be overlooked that BT, like any major telecommunications company, is capital intensive. Commonly interest on capital and provision for depreciation can account for more than one third of the annual expenditure of a telecommunications company. Certainly in the short term ways of raising manpower productivity must constantly be looked for, but in the long term it is greater asset-utilization that will determine the rate of return.

PRIVATIZATION

In 1984 shares of British Telecommunications were marketed for the first time and the response of investors was overwhelming. It is too early to comment on the likely effects on the future of the corporation. An indication has been given above of the complexity of the financial issues the board faces in meeting the challenge of competition. BT continues to employ more than a quarter of a million staff and possibly the way in which they are organized, managed, and motivated will be as important as any other factor in determining BT's future. Privatization in itself is probably less important than the freedom given to others to compete in the marketing of telecommunications terminals and the transmission of telecommunications.

In the United States the problems facing AT & T and the local Bell Telephone companies are very similar to those which BT faces, albeit from a different statutory approach. One possible development which, undoubtedly like AT & T, the board of British Telecommunications will be considering will be diversification. BT has considerable strengths in telecommunications expertise. It has a long history of being in the forefront of telecommunications research and development. It clearly has the potential to enter overseas markets as a means of generating more income to sustain its home-based services.

VII

CONCLUSION

32

FUTURE TRENDS

It was argued that the world is on the threshold of a technological revolution and that telecommunications had a major part to play in reaping the benefits for mankind. Undoubtedly telecommunications technology will continue to advance rapidly. In the voice field, telephone terminals will become smaller and fully mobile. It is inconceivable that voice telecommunication will continue to depend on being tied to the wall or floor by flexible connector, or that instant communication, wherever the user is, will not become commonplace. In time the artificial identity of users by means of numbering systems will become obsolescent and connections will be established by voice-operated selection equipment.

Terminals will be used for both voice and nonvoice telecommunications, and will incorporate software to carry out a wide range of functions. Some form of visual indication will be given of the information being communicated. Speeds of response will be as fast as the user can keep up with. Reliability and security of telecommunications will continue to improve, and system failures will be bypassed automatically.

In time the charges for telecommunications will become more broadly based and less related to the amount of use. Charges are likely to be based on the capacity and range of services being provided and less on the precise time they are used. Inevitably, given telecommunications' importance to the political, industrial, commercial, and social life of the nation they will continue to be closely regulated by the State, whatever the basis of ownership of the systems.

33

PERSPECTIVE

In this book the technicalities and engineering of telecommunications have deliberately been discussed in no greater depth than was necessary for an understanding of the direction and management required to ensure that systems perform efficiently and are cost effective. The advances made in telecommunications technology and the speed of technical change are truly remarkable, requiring degrees of precision which, a generation ago, would have been unbelievable. The progress achieved is even more remarkable when it is borne in mind that the users of telecommunications are given negligible training and instruction, and that users are at times incompetent and intolerant.

If, however, the potential of telecommunications technology is to be realized to the full, it is imperative that business managers and directors understand that they have an increasingly important part to play and that they can no longer leave major critical decisions which affect the operation and viability of their concerns to the long-established telecommunications companies.

Inescapably the perspective of this book stems from the author's career which spans more than forty years in the British Post Office and later, British Telecommunications. His close working association over the years with his opposite numbers in the United States, Europe, and the Commonwealth, leaves him in no doubt that the fundamentals of business telecommunications management apply universally.

APPENDICES

UNITED KINGDOM GROWTH OF TELEPHONES IN SERVICE

Year ending 31 March	Total telephones in service (000s)	Increase over 10 years (000s)
1920	915	—
1930	1896	981
1940	3339	1443
1950	5171	1832
1960	7856	2685
1970	13959	6103
1980	26737	12778

Increase 1920–1970 13,044,000
Increase 1970–1980 12,778,000

UNITED KINGDOM GROWTH OF INLAND TELEPHONE CALLS

Year Ending 31 March	Total telephone calls (000,000s)	Increase over 10 years (000,000s)
1920	849	—

UNITED KINGDOM GROWTH OF INLAND TELEPHONE CALLS (*continued*)

Year Ending 31 March	Total telephone calls (000,000s)	Increase over 10 years (000,000s)
1930	1322	473
1940	2215	893
1950	3175	960
1960	4287	1112
1970	9622	5335
1980	19857	10235

Increase 1920–1970 .... 8,773 millions
Increase 1970–1980 .... 10,235 millions

UNITED KINGDOM INVESTMENT IN LOCAL TELEPHONE EXCHANGES

Year ending 31 March	Type of exchange				
	Manual	Strowger	Crossbar	Electronic	Total
1920	3262	13	—	—	3275
1930	4323	307	—	—	4630
1940	2510	3213	—	—	5723
1950	1775	4091	—	—	5866
1960	921	5088	—	—	6099
1970	105	5991	11	31	6138
1980	—	4552	559	1159	6260

Percentage of exchange lines	1970	1980
Served by manual exchanges	1.4	—
Served by Stowger exchanges	97.9	72.9
Served by crossbar exchanges	0.3	16.0
Served by electronic exchanges	0.4	11.1

UNITED KINGDOM INVESTMENT IN TELECOMMUNICATIONS DISTRIBUTION

Year ending 31 March	Local distribution cable pairs (000s)	Main network public speech channels
1920	—	1707
1930	—	3680
1940	—	6220
1950	4605	16340
1960	6942	27195
1970	12848	87517
1980	24864	221569

UNITED KINGDOM TELECOMMUNICATIONS ANNUAL INCOME AND EXPENDITURE

Year ending 31 March	Expenditure (£000s)	Income (£000s)	Balance (£000s)
1920	11,193	9,231	−1,962
1930	21,379	21,892	513
1940	37,920	38,914	995
1950	79,993	89,126	9,133
1960	191,569	208,496	16,927
1970	536,046	588,131	52,085
1980	3,429,800	3,558,900	129,100

Note: Expenditure includes provision for depreciation and interest on outstanding capital.

UNITED KINGDOM NET BOOK VALUE OF FIXED ASSETS

	Year ending 31 March	
	1970	1980
Fixed assets per exchange line	£	£
Inter-exchange circuits		57.4
Local distribution		64.2
Customers' terminals		57.7
Exchange equipment		133.9
Remainder		8.8
Total inland plant		322.0
International services		13.4
Accommodation		57.7
Miscellaneous		22.1
Total	264.9	415.4
Fixed assets per telephone call	0.235	0.366
Exchange connections per employee	37	73
Telephone calls per employee (000s)	42	83

TELECOMMUNICATIONS OPERATIONAL EXPENDITURE

The following figures are illustrative. The percentages will vary from one company to another, depending in part on allocation of expenditure to particular operational accounting sub-heads.

Operational sub-head	Percentage of income
Maintenance	13
Operator services	8
Accommodation	7
Marketing & sales	2
Billing	2
Motor transport	2

INDEX

Access control, and protection of sensitive and confidential
        information, 90
Accommodation costs, in financial appraisal of technical options, 117
Accountability, operational, resource management and, 153–155
American Telephone & Telegraph Company (AT & T), implications of
        privatization for, 177
Analog signals, in local networks transmission, 41–42
Availability
    as criterion for installation standards, 67
    as criterion for service standards, 71–78

Banks, data transmission in, 33
Bell, Alexander Graham, 8, 11
Billing
    accuracy of, as standard of service, 99–100
    in future, 181
British Telecommunications
    in cellular radio network competition, 172–173
    in terminal market competition, 169–172
    privatization and, 177
    value added network services regulation involving, 173–174

Cable and Wireless Limited, 164
Capital expenditures, in financial appraisal of technical options, 117
Cellular radio networks. *See* Networks, cellular radio

Coaxial cables
    in construction of main networks, 53, 114
    in local networks transmission, 42
Coding, as method of protecting sensitive or confidential information, 90
Communications
    data
        applications in business and industry, 20–21, 32–34
        terminals for, 32–34
    history and development, 3–6, 7–10
    message
        terminals for, 29–31
    voice
        in future, 181
        terminals for, 25–29
Competition, regulation and, 169–174
Computers, applications in telecommunications technology, 11–12, 32
Connections
    quality of, as standard of service, 81–82
    speed of, as standard of service, 79–80
Cost
    as criterion for installation standards, 67
    in development-procurement plans, 117

Data communications. See Communications, data
Depreciation, in financial appraisal of technical options, 117
Development, technological, as standard of service, 101–102
Development-procurement plans. See plans, development-procurement
Dialed services, management of, 133–137
Digital techniques, in main network transmission, 54–55
Directories, as source of number information, 92
Directory inquiry services, and provision of number information, 92
Drums, as coded communication signal, 7

Edison Telephone Company (London), 163
Efficiency, as criterion for installation standards, 67
Electronic mail, in message communication, 3
Employment levels, in utilization plans, 122
Enhancement
    in financial appraisal of technical options, 117
    technological, as standard of service, 101–102
Equipment allocation, as function of switching systems, 49

Equipment utilization, resource management and, 145–147
Expansion, in financial appraisal of technical options, 117
Expenditures, in utilization plans, 122

Facsimile, as form of message communication, 30
Financial analysis, in development and procurement planning, 109,
     116–117
Financial performance. *See* Performance, financial
Fire, as communications medium, 7
Flags, as form of visual communication, 7
Forecasting, in development and procurement planning, 109, 111–113

Heliographs, as form of visual communication, 7

Income levels, in utilization plans, 122
Industrial relations, resource management and, 150–152
Industrial Revolution
   as consequence of development of printing press, 3
   and development of communications, 4
Installation
   factors in achieving optimal speed of, 63–66
   standardization of, 67–69
Interfaces, user-system
   in future visual displays, 17–18
   in telegraph, 17
   in telephone, 15–16
International calls, terminals for, 27

Labor unions, and industrial relations for resource management, 151
Leasing, decision making associated with, 117
Leasing costs, in financial appraisal of technical options, 117
Liberalization
   as consequence of regulation, 169–171
   financial implications of, 174–177
Light, as communications medium, 7
Load assessment, in utilization planning, 121
Local networks. *See* Networks, local

Main networks. *See* Networks, main
Maintenance, and achievement of standards of service, 133–134
Maintenance costs, in financial appraisal of technical options, 117

Management
  decision making and, 12–13
  operational
    and achievement of standards of service, 127–143
    responsibilities of, 125, 126
    of resources, 145–155
  strategic, responsibilities of, 125
Manpower productivity, resource management and, 147–150
Mercury, in competition with British Telecommunications,
      171–172
Message systems, teminals for, 29–31
Monopoly, and freedom of user choice, 159–161
Morse code, 29, 30

National Telephone Company (Great Britain), 163, 164
Networks
  cellular radio, regulation and, 172–173
  local
    data transmission in, 41–42
    external, 42–43
    internal, 39–41
  main
    design and characteristics, 51, 52
    methods of constructing and developing, 52–55
    and standards of service, 76
  public access to, 36
  value added
    and competition, 173–174
    services and leasing options involving, 55–56
Number information, provision of, as support service, 91–94

Operator-assisted calls, 74
Operators, role in achievement of standards of service,
      127–132
Opinion polling, in development of installation standards, 68
Optical fibers, in construction of main networks, 55, 115
Overhead wires, in construction of main networks, 52–53

Parliament, regulation of telecommunications industry by,
      164–165, 166–167

Pay phones, as support service, 95
Performance
    as criterion for installation standards, 67, 69
    financial, resource management and, 152–153
    standards of, in utilization plans, 121
Planning
    development and procurement
        definition, 106
        financial analysis in, 109, 116–117
        forecasting in, 109, 111–113
        identification of impinging factors in, 109, 115–116
        identification of technical options in, 109, 113–115
        usage assessment in, 109–111
    nature of, 105–106
    strategic
        aims, 107–108
        definition, 106
    utilization, definition, 106
    See also Plans, development and procurement; Plans, utilization
Plans
    development and procurement
        sensitivity analysis of, 118–120
        steps in preparation of, 117–118
    utilization, steps in preparation of, 121–122
    See also Planning
Post Office
    historical background of control of British telecommunications by,
        163, 164
    Parliamentary regulation of, 164–166
    as public corporation, 166–168
Power and heating costs, and financial appraisal of technical options,
        117
Private branch exchanges, 49
Privatization, as consequence of regulation, 177
Promptness
    as criterion for installation standards, 63–66, 67–68
    as criterion for service standards, 79–80, 85–86
Productivity, manpower. See Manpower productivity
Pulse code modulation, in local networks transmission, 42
Purchasing, outright, decision making associated with, 117

Radio, in construction of main networks, 53, 115

Record keeping
   as factor in achievement of optimal speed of repair, 85, 86, 87
   in management of dialed services, 135, 136

Regulation
   of cellular radio networks, 172–173
   as factor leading to competition and liberalization, 169–171
   as factor leading to privatization, 177
   history, 163–168
   and impinging factor in development and procurement planning, 115
   of value added network services, 173–174

Reliability, as standard of service, 83–84

Remote control, terminals for, 35

Renaissance, and development of printing press, 3

Repair services
   management and achievement of standards in, 137–139
   speed of, as standard of performance, 85–87

Repeat call facilities, 27

Resource management
   equipment utilization and, 145–147
   financial performance and, 152–153
   industrial relations and, 150–152
   manpower productivity and, 147–150
   operational accountability and, 153–155

Satellites, in construction of main networks, 54, 115

Security, as standard of service, 89–90

Sensitivity analysis, of development and procurement plans, 118–120

Service, standards of
   applications to installation, 63–66, 67–69
   availability of communications and, 71–78
   billing accuracy and, 99–100
   implications for technological enhancement and development, 101–102
   management and achievement of, 139–142
   methods of achieving
      for dialed services, 133–137
      for installation repair services, 137–139
      for operator services, 127–172
      for service provision, 139–142
   quality of connection and, 81–82
   reliability and, 83–84

Service, standards of, (*cont.*)
  security and, 89–90
  speed of connection and, 79–80
  speed of repair and, 85–87
  support services and, 91–95
  user relations and, 97–98
Servicing costs, in financial assessment of technical options, 117
Smoke, as communication medium, 7
Standardization
  of installation, 67–69
  of performance, 121
  of service, 59–102
Statistics, and measurement of service standards, 134
Support services
  management and achievement of standards, 142–143
  and standardization of service, 91–95
Surveillance, data terminals for, 34
Switchboards, and standards of service, 72–73, 130–131
Switching equipment, and standards of service, 73–76
Switching systems
  active, 48
  common user, 49
  dedicated, 49
  electronic, and measurement of service standards, 135
  functions, 49
  methods of selecting extension numbers for, 45–47
  passive, 47
System costs, redeployment of, in financial appraisal of technical options,
    117

Technological change
  future consequences of, 181
  as impinging factor in development and procurement planning, 115–116
Telegraph, applications, 8
Telemetry, terminals for, 35
Telephone
  as basic terminal for voice communications, 25–26
  communications technology involving, 8–9
  impact on industry and commerce, 11
Teleprinters, in public telegraph services, 30
Tele-text services, applications, 35–36

Telex, in message communications, 30–31
Terminal identification, as function of switching systems, 49
Terminals
   for data communications, 32–34
   factors delaying or hindering installation of, 65–66
   future design and applications of, 181
   for message communications, 29–31
   for surveillance, 34
   for telemetry and remote control, 35
   for voice communications, 25–29
Transmission quality, as standard of dialed service, 136

Unemployment, association of technological change with, 115–116
"Unfortunate" calls, prevention of, 128–129, 130
United Kingdom
   annual expenditures and income for telecommunications (1920–1980),
      187
   growth in number of inland telephone calls (1920–1980), 185
   growth of telephone services (1920–1980), 185
   investment in local telephone exchanges in (1920–1980), 186–187
   investment in telecommunications distribution (1920–1980), 187
   net book value of fixed telecommunications assets (1970 and 1980), 188
   regulation of telecommunications in, 163–177
   telecommunications operational expenditures, 188
United States, telecommunications privatization in, 177
Usage assessment
   in development and procurement planning, 109–111
   in utilization planning, 121–122
User assistance, as support service, 94–95
User notification, as method of providing number information, 92
User reports, in evaluation of service, 85
Users, and standards of service, 97–98
Utilization plans. See Plans, utilization

Value added networks. See Networks, value added
Visual displays
   in telephone communication, 27–28
   and user-system interfaces, 17
Voice communications. See Communications, voice

Word processing, in message communication, 31